Python Programming for Data Analysis

José Unpingco

Python Programming
for Data Analysis

 Springer

José Unpingco
University of California
San Diego
CA, USA

ISBN 978-3-030-68954-4   ISBN 978-3-030-68952-0   (eBook)
https://doi.org/10.1007/978-3-030-68952-0

This Springer imprint is published by the registered company Springer Nature Switzerland AG
The registered company address is: Gewerbestrasse 11, 6330 Cham, Switzerland

To Irene, Nicholas, and Daniella, for all their patient support.

Preface

This book grew out of notes for the ECE143 Programming for Data Analysis class that I have been teaching at the University of California, San Diego, which is a requirement for both graduate and undergraduate degrees in Machine Learning and Data Science. The reader is assumed to have some basic programming knowledge and experience using another language, such as Matlab or Java. The Python idioms and methods we discuss here focus on data analysis, notwithstanding Python's usage across many other topics. Specifically, because raw data is typically a mess and requires much work to prepare, this text focuses on specific Python language features to facilitate such cleanup, as opposed to only focusing on particular Python modules for this.

As with ECE143, here we discuss why things are the way they are in Python instead of just that they are this way. I have found that providing this kind of context helps students make better engineering design choices in their codes, which is especially helpful for newcomers to both Python and data analysis. The text is sprinkled with little tricks of the trade to make it easier to create readable and maintainable code suitable for use in both production and development.

The text focuses on using the Python language itself effectively and then moves on to key third-party modules. This approach improves effectiveness in different environments, which may or may not have access to such third-party modules. The Numpy numerical array module is covered in depth because it is the foundation of all data science and machine learning in Python. We discuss the Numpy array data structure in detail, especially its memory aspects. Next, we move on to Pandas and develop its many features for effective and fluid data processing. Because data visualization is key to data science and machine learning, third-party modules such as Matplotlib are developed in depth, as well as web-based modules such as Bokeh, Holoviews, Plotly, and Altair.

On the other hand, I would *not* recommend this book to someone with no programming experience at all, but if you can do a little Python already and want to improve by understanding how and why Python works the way it does, then this is a good book for you.

To get the most out of this book, open a Python interpreter and type-along with the many code samples. I worked hard to ensure that all of the given code samples work as advertised.

Acknowledgments I would like to acknowledge the help of Brian Granger and Fernando Perez, two of the originators of the Jupyter Notebook, for all their great work, as well as the Python community as a whole, for all their contributions that made this book possible. Hans Petter Langtangen was the author of the Doconce [1] document preparation system that was used to write this text. Thanks to Geoffrey Poore [2] for his work with PythonTeX and LaTeX, both key technologies used to produce this book.

San Diego, CA, USA
February, 2020

José Unpingco

References

1. H.P. Langtangen, DocOnce markup language. https://github.com/hplgit/doconce
2. G.M. Poore, Pythontex: reproducible documents with latex, python, and more. Comput. Sci. Discov. **8**(1), 014010 (2015)

Contents

Chapter 1
Basic Programming

1.1 Basic Language

Before we get into the details, it is a good idea to get a high-level orientation to Python. This will improve your decision-making later regarding software development for your own projects, especially as these get bigger and more complex. Python grew out of a language called ABC, which was developed in the Netherlands in the 1980s as an alternative to BASIC to get scientists to utilize microcomputers, which were new at the time. The important impulse was to make non-specialist scientists able to productively utilize these new computers. Indeed, this pragmatic approach continues to this day in Python which is a direct descendent of the ABC language.

There is a saying in Python—*come for the language, stay for the community.* Python is an open source project that is community driven so there is no corporate business entity making top-down decisions for the language. It would seem that such an approach would lead to chaos but Python has benefited over many years from the patient and pragmatic leadership of Guido van Rossum, the originator of the language. Nowadays, there is a separate governance committee that has taken over this role since Guido's retirement. The open design of the language and the quality of the source code has made it possible for Python to enjoy many contributions from talented developers all over the world over many years, as embodied by the richness of the standard library. Python is also legendary for having a welcoming community for newcomers so it is easy to find help online for getting started with Python.

The pragmatism of the language and the generosity of the community have long made Python a great way to develop web applications. Before the advent of data science and machine learning, upwards of 80% of the Python community were web developers. In the last five years (at the time of this writing), the balance is tilted to an almost even split between web developers and data scientists. This is the reason you find a lot of web protocols and technology in the standard library.

Python is an interpreted language as opposed to a compiled language like C or FORTRAN. Although both cases start with a source code file, the compiler

J. Unpingco, *Python Programming for Data Analysis*, https://doi.org/10.1007/978-3-030-68952-0_1

examines the source code end-to-end and produces an executable that is linked to system-specific library files. Once the executable is created, there is no longer any need for the compiler. You can just run the executable on the system. On the other hand, an interpreted language like Python you must always have a running Python process to execute the code. This is because the Python process is an abstraction on the platform it is running on and thus must interpret the instructions in the source code to execute them on the platform. As the intermediary between the source code on the platform, the Python interpreter is responsible for the platform specific issues. The advantage of this is that source code can be run on different platforms as long as there is a working Python interpreter on each platform. This makes Python source codes portable between platforms because the platform specific details are handled by the respective Python interpreters. Portability between platforms was a key advantage of Python, especially in the early days. Going back to compiled languages, because the platform specific details are embedded in the executable, the executable is tied to a specific platform and to those specific libraries that have been linked into the executable. This makes these codes are less portable than Python, but because the compiler is able to link to the specific platform it has the option to exploit platform- specific level optimizations and libraries. Furthermore, the compiler can study the source code file and apply compiler-level optimizations that accelerate the resulting executable. In broad strokes, those are the key differences between interpreted and compiled languages. We will later see that there are many compromises between these two extremes for accelerating Python codes.

It is sometimes said that Python is *slow* as compared to compiled languages, and pursuant to the differences we discussed above, that may be expected, but it is really a question of where the clock starts. If you start the clock to account for developer time, not just code runtime, then Python is clearly faster, just because the development iteration cycle does not require a tedious compile and link step. Furthermore, Python is just simpler to use than compiled languages because so many tricky elements like memory management are handled automatically. Pythons quick turnaround time is a major advantage for product development, which requires rapid iteration. On the other hand, codes that are compute-limited and must run on specialized hardware are not good use-cases for Python. These include solving systems of parallel differential equations simulating large-scale fluid mechanics or other large-scale physics calculations. Note that Python *is* used in such settings but mainly for staging these computations or postprocessing the resulting data.

1.1.1 Getting Started

Your primary interface to the Python interpreter is the commandline. You can type in python in your terminal you should see something like the following,

```
Python 3.7.3 (default, Mar 27 2019, 22:11:17)
[GCC 7.3.0] :: Anaconda, Inc. on linux
```

```
Type "help", "copyright", "credits" or "license" for more
↪   information.
>>>
```

There is a lot of useful information including the version of Python and its provenance. This matters because sometimes the Python interpreter is compiled to allow fast access to certain preferred modules (i.e., `math` module). We will discuss this more when we talk about Python modules.

1.1.2   Reserved Keywords

Although Python will not stop you, do not use these as variable or function names.

```
and        del        from       not        while
as         elif       global     or         with
assert     else       if         pass       yield
break      except     import     print
class      exec       in         raise
continue   finally    is         return
def        for        lambda     try
```

nor these neither

```
abs all any ascii bin bool breakpoint bytearray bytes callable
chr classmethod compile complex copyright credits delattr
dict dir display divmod enumerate eval exec filter float
format frozenset getattr globals hasattr hash help hex id
input int isinstance issubclass iter len list locals map max
memoryview min next object oct open ord pow print property
range repr reversed round set setattr slice sorted
staticmethod str sum super tuple type vars zip
```

For example, a common mistake is to assign `sum=10`, not realizing that now the `sum()` Python function is no longer available.

1.1.3   Numbers

Python has common-sense number-handling capabilities. The comment character is the hash (#) symbol.

```
>>> 2+2
4
>>> 2+2   # and a comment on the same line as code
4
>>> (50-5*6)/4
5.0
```

Note that division in Python 2 is integer-division and in Python 3 it is floating point division with the `//` symbol providing integer-division. Python is *dynamically*

typed, so we can do the following assignments without declaring the types of
`width` and `height`.

```
>>> width = 20
>>> height = 5*9
>>> width * height
900
>>> x = y = z = 0   # assign x, y and z to zero
>>> x
0
>>> y
0
>>> z
0
>>> 3 * 3.75 / 1.5
7.5
>>> 7.0 / 2 # float numerator
3.5
>>> 7/2
3.5
>>> 7 // 2 # double slash gives integer division
3
```

It is best to think of assignments as labeling values in memory. Thus, `width` is a
label for the number `20`. This will make more sense later.[1] Since Python 3.8, the
walrus assignment operator allows the assignment itself to have the value of the
assignee, as in the following,

```
>>> print(x:=10)
10
>>> print(x)
10
```

The operator has many other subtle uses and was introduced to improve readability
in certain situations. You can also cast among the numeric types in a common-sense
way:

```
>>> int(1.33333)
1
>>> float(1)
1.0
>>> type(1)
<class 'int'>
>>> type(float(1))
<class 'float'>
```

Importantly, Python integers are of arbitrary length and can handle extremely large
integers. This is because they are stored as a list of digits internally. This means that
they are slower to manipulate than Numpy integers which have fixed bit-lengths,
provided that the integer can fit into allocated memory.

---

[1] Note http://www.pythontutor.com is a great resource for exploring how variables are assigned in
Python.

1.1.4   Complex Numbers

Python has rudimentary support for complex numbers.

```
>>> 1j * 1J
(-1+0j)
>>> 1j * complex(0,1)
(-1+0j)
>>> 3+1j*3
(3+3j)
>>> (3+1j)*3
(9+3j)
>>> (1+2j)/(1+1j)
(1.5+0.5j)
>>> a=1.5+0.5j
>>> a.real # the dot notation gets an attribute
1.5
>>> a.imag
0.5
>>> a=3.0+4.0j
>>> float(a)
Traceback (most recent call last):
  File "<stdin>", line 1, in <module>
TypeError: can't convert complex to float
>>> a.real
3.0
>>> a.imag
4.0
>>> abs(a)   # sqrt(a.real**2 + a.imag**2)
5.0
>>> tax = 12.5 / 100
>>> price = 100.50
>>> price * tax
12.5625
>>> price + _
113.0625
>>> # the underscore character is the last evaluated result
>>> round(_, 2) # the underscore character is the last evaluated
↪   result
113.06
```

We typically use Numpy for complex numbers, though.

1.1.5   Strings

String-handling is very well developed and highly optimized in Python. We just
cover the main points here. First, single or double quotes define a string and there is
no precedence relationship between single and double quotes.

```
>>> 'spam eggs'
'spam eggs'
>>> 'doesn\'t' # backslash defends single quote
"doesn't"
>>> "doesn't"
"doesn't"
>>> '"Yes," he said.'
'"Yes," he said.'
>>> "\"Yes,\" he said."
'"Yes," he said.'
>>> '"Isn\'t," she said.'
'"Isn\'t," she said.'
```

Python strings have C-style escape characters for newlinewsnewlines, tabs, etc.
String literals defined this way are by default encoded using UTF-8 instead of
ASCII, as in Python 2. The triple single (') or triple double quote (") denotes a
block with embedded newlines or other quotes. This is particularly important for
function documentation docstrings that we will discussed later.

```
>>> print( '''Usage: thingy [OPTIONS]
... and more lines
... here and
... here
...         ''')
Usage: thingy [OPTIONS]
and more lines
here and
here
```

Strings can be controlled with a character before the single or double quote. For
example, see the comments (#) below for each step,

```
>>> # the 'r' makes this a 'raw' string
>>> hello = r"This long string contains newline characters \n, as
↪   in C"
>>> print(hello)
This long string contains newline characters \n, as in C
>>> # otherwise, you get the newline \n acting as such
>>> hello = "This long string contains newline characters \n, as
↪   in C"
>>> print(hello)
This long string contains newline characters
, as in C
```

```
>>> u'this a unicode string  μ ±' # the 'u' makes it a unicode
↪    string for Python2
'this a unicode string  μ ±'
>>> 'this a unicode string  μ ±'  # no 'u' in Python3 is still
↪    unicode string
'this a unicode string  μ ±'
>>> u'this a unicode string \xb5 \xb1' # using hex-codes
'this a unicode string μ ±'
```

Note that a `f-string` evaluates (i.e., *interpolates*) the Python variables in the current scope,

```
>>> x = 10
>>> s = f'{x}'
>>> type(s)
<class 'str'>
>>> s
'10'
```

Beware that an f-string is not resolved until run-time because it has to resolve the embedded variables. This means that you *cannot* use f-strings as docstrings. Importantly, Python strings are *immutable* which means that once a string is created, it *cannot* be changed in-place. For example,

```
>>> x = 'strings are immutable '
>>> x[0] = 'S' # not allowed!
Traceback (most recent call last):
   File "<stdin>", line 1, in <module>
TypeError: 'str' object does not support item assignment
```

This means you have to create new strings to make this kind of change.

Strings vs Bytes In Python 3, the default string encoding for string literals is UTF-8. The main thing to keep in mind is that bytes and strings are now distinct objects, as opposed to both deriving from `basestring` in Python 2. For example, given the following unicode string,

```
>>> x='Ø'
>>> isinstance(x,str)    # True
True
>>> isinstance(x,bytes)  # False
False
>>> x.encode('utf8')     # convert to bytes with encode
b'\xc3\x98'
```

Note the distinction between bytes and strings. We can convert bytes to strings using `decode`,

```
>>> x=b'\xc3\x98'
>>> isinstance(x,bytes)  # True
True
>>> isinstance(x,str)    # False
False
>>> x.decode('utf8')
'Ø'
```

An important consequence is that you cannot append strings and bytes as in the
following: u"hello"+b"goodbye". This used to work fine in Python 2 because
bytes would automatically be decoded using ASCII, but this no longer works in
Python 3. To get this behavior, you have to explicitly decode/encode. For
example,

```
>>> x=b'\xc3\x98'
>>> isinstance(x,bytes) # True
True
>>> y='banana'
>>> isinstance(y,str)    # True
True
>>> x+y.encode()
b'\xc3\x98banana'
>>> x.decode()+y
'Øbanana'
```

Slicing Strings Python is a zero-indexed language (like C). The colon (:) character
denotes .

```
>>> word = 'Help' + 'A'
>>> word
'HelpA'
>>> '<' + word*5 + '>'
'<HelpAHelpAHelpAHelpAHelpA>'
>>> word[4]
'A'
>>> word[0:2]
'He'
>>> word[2:4]
'lp'
>>> word[-1]       # The last character
'A'
>>> word[-2]       # The last-but-one character
'p'
>>> word[-2:]      # The last two characters
'pA'
>>> word[:-2]      # Everything except the last two characters
'Hel'
```

String Operations Some basic numerical operations work with strings.

```
>>> 'hey '+'you' # concatenate with plus operator
'hey you'
>>> 'hey '*3       # integer multiplication duplicates strings
'hey hey hey '
>>> ('hey ' 'you') # using parentheses without separating comma
'hey you'
```

Python has a built-in and very powerful regular expression module (re) for string
manipulations. String substitution creates new strings.

```
>>> x = 'This is a string'
>>> x.replace('string','newstring')
'This is a newstring'
>>> x # x hasn't changed
'This is a string'
```

Formatting Strings There are *so* many ways to format strings in Python, but here is the simplest that follows C-language `sprintf` conventions in conjunction with the modulo operator `%`.

```
>>> 'this is a decimal number %d'%(10)
'this is a decimal number 10'
>>> 'this is a float %3.2f'%(10.33)
'this is a float 10.33'
>>> x = 1.03
>>> 'this is a variable %e' % (x) # exponential format
'this is a variable 1.030000e+00'
```

Alternatively, you can just join them using +,

```
>>> x = 10
>>> 'The value of x = '+str(x)
'The value of x = 10'
```

You can format using dictionaries as in the following,

```
>>> data = {'x': 10, 'y':20.3}
>>> 'The value of x = %(x)d and y = %(y)f'%(data)
'The value of x = 10 and y = 20.300000'
```

You can use the `format` method on the string,

```
>>> x = 10
>>> y = 20
>>> 'x = {0}, y = {1}'.format(x,y)
'x = 10, y = 20'
```

The advantage of `format` is you can reuse the placeholders as in the following,

```
>>> 'x = {0},{1},{0}; y = {1}'.format(x,y)
'x = 10,20,10; y = 20'
```

And also the f-string method we discussed above.

Programming Tip: Python 2 Strings

In Python 2, the default string encoding was 7-bit ASCII. There was no distinction between bytes and strings. For example, you could read from a binary-encoded JPG file as in the following,

```
with open('hour_1a.jpg','r') as f:
    x = f.read()
```

This works fine in Python 2 but throws a `UnicodeDecodeError` error in Python 3. To fix this in Python 3, you have to read using the `rb` binary mode instead of just the `r` file mode.

Basic Data Structures Python provides many powerful data structures. The two most powerful and fundamental are the list and dictionary. Data structures and algorithms go hand-in-hand. If you do not understand data structures, then you cannot effectively write algorithms and vice versa. Fundamentally, data structures provide guarantees to the programmer that will be fulfilled if the data structures are used in the agreed-upon manner. These guarantees are known as the *invariants* for the data structure.

Lists The list is an order-preserving general container that implements the *sequence* data structure. The invariant for the list is that indexing a non-empty list will always give you the next valid element in *order*. Indeed, the list is Python's primary ordered data structure. This means that if you have a problem where order is important, then you should be thinking about the list data structure. This will make sense with following examples.

```
>>> mix = [3,'tree',5.678,[8,4,2]]  # can contain sublists
>>> mix
[3, 'tree', 5.678, [8, 4, 2]]
>>> mix[0]      # zero-indexed Python
3
>>> mix[1]      # indexing individual elements
'tree'
>>> mix[-2]     # indexing from the right, same as strings
5.678
>>> mix[3]      # get sublist
[8, 4, 2]
>>> mix[3][1]   # last element is sublist
4
>>> mix[0] = 666 # lists are mutable
>>> mix
[666, 'tree', 5.678, [8, 4, 2]]
>>> submix = mix[0:3] # creating sublist
>>> submix
[666, 'tree', 5.678]
>>> switch = mix[3] + submix # append two lists with plus
>>> switch
[8, 4, 2, 666, 'tree', 5.678]
>>> len(switch)  # length of list is built-in function
6
>>> resize=[6.45,'SOFIA',3,8.2E6,15,14]
>>> len(resize)
6
>>> resize[1:4] = [55]   # assign slices
>>> resize
[6.45, 55, 15, 14]
>>> len(resize) # shrink a sublist
4
>>> resize[3]=['all','for','one']
>>> resize
[6.45, 55, 15, ['all', 'for', 'one']]
>>> len(resize)
4
>>> resize[4]=2.87  # cannot append this way!
Traceback (most recent call last):
  File "<stdin>", line 1, in <module>
IndexError: list assignment index out of range
>>> temp = resize[:3]
```

```
>>> resize = resize + [2.87] # add to list
>>> resize
[6.45, 55, 15, ['all', 'for', 'one'], 2.87]
>>> len(resize)
5
>>> del resize[3]      # delete item
>>> resize
[6.45, 55, 15, 2.87]
>>> len(resize)        # shorter now
4
>>> del resize[1:3]   # delete a sublist
>>> resize
[6.45, 2.87]
>>> len(resize)        # shorter now
2
```

Programming Tip: Sorting Lists
The built-in function `sorted` sorts lists,

```
>>> sorted([1,9,8,2])
[1, 2, 8, 9]
```

Lists can also be sorted in-place using the `sort()` list method,

```
>>> x = [1,9,8,2]
>>> x.sort()
>>> x
[1, 2, 8, 9]
```

Both of these use the powerful Timsort algorithm. Later, we will see more variations and uses for these sorting functions.

Now that we have a feel for how to index and use lists, let us talk about the invariant that it provides: as long as you index a list within its bounds, it provides the next ordered element of the list. For example,

```
>>> x = ['a',10,'c']
>>> x[1] # return 10
10
>>> x.remove(10)
>>> x[1] # next element
'c'
```

Notice that the list data structure filled in the gap after removing 10. This is extra work that the list data structure did for you without explicitly programming. Also, list elements are accessible via integer indices and integers have a natural ordering and thus so does the list. The work of maintaining the invariant does not come for free, however. Consider the following,

```
>>> x = [1,3,'a']
>>> x.insert(0,10) # insert at beginning
```

```
>>> x
[10, 1, 3, 'a']
```

Seem harmless? Sure, for small lists, but not so for large lists. This is because to maintain the invariant the list has to scoot (i.e., memory copy) the remaining elements over to the right to accommodate the new element added at the beginning. Over a large list with millions of elements and in a loop, this can take a substantial amount of time. This is why the default `append()` and `pop()` list methods work at the tail end of the list, where there is no need to scoot items to the right.

Tuples Tuples are another general purpose sequential container in Python, very similar to lists, but these are *immutable*. Tuples are delimited by commas (parentheses are grouping symbols). Here are some examples,

```
>>> a = 1,2,3 # no parenthesis needed!
>>> type(a)
<class 'tuple'>
>>> pets=('dog','cat','bird')
>>> pets[0]
'dog'
>>> pets + pets # addition
('dog', 'cat', 'bird', 'dog', 'cat', 'bird')
>>> pets*3
('dog', 'cat', 'bird', 'dog', 'cat', 'bird', 'dog', 'cat', 'bird')
>>> pets[0]='rat' # assignment not work!
Traceback (most recent call last):
  File "<stdin>", line 1, in <module>
TypeError: 'tuple' object does not support item assignment
```

It may seem redundant to have tuples which behave in terms of their indexing like lists, but the key difference is that tuples are immutable, as the last line above shows. The key advantage of immutability is that it comes with less overhead for Python memory management. In this sense, they are lighter weight and provide stability for codes that pass tuples around. Later, we will see this for function signatures, which is where the major advantages of tuples arise.

Programming Tip: Understanding List Memory

Python's `id` function shows an integer corresponding to the internal reference for a given variable. Earlier, we suggested considering variable assignment as labeling because internally Python works with a variable's `id`, not its variable name/label.

```
>>> x = y = z = 10.1100
>>> id(x)  # different labels for same id
140271927806352
>>> id(y)
140271927806352
>>> id(z)
140271927806352
```

(continued)

This is more consequential for mutable data structures like lists. Consider the following,

```
>>> x = y = [1,3,4]
>>> x[0] = 'a'
>>> x
['a', 3, 4]
>>> y
['a', 3, 4]
>>> id(x),id(y)
(140271930505344, 140271930505344)
```

Because x and y are merely two labels for the same underlying list, changes to one of the labels affects *both* lists. Python is inherently stingy about allocating new memory so if you want to have two different lists with the same content, you can force a copy as in the following,

```
>>> x = [1,3,4]
>>> y = x[:]      # force copy
>>> id(x),id(y) # different ids now!
(140271930004160, 140271929640448)
>>> x[1] = 99
>>> x
[1, 99, 4]
>>> y # retains original data
[1, 3, 4]
```

Tuple Unpacking Tuples unpack assignments in order as follows:,

```
>>> a,b,c = 1,2,3
>>> a
1
>>> b
2
>>> c
3
```

Python 3 can unpack tuples in chunks using the * operator,

```
>>> x,y,*z  = 1,2,3,4,5
>>> x
1
>>> y
2
>>> z
[3, 4, 5]
```

Note how the z variable collected the remaining items in the assignment. You can also change the order of the chunking,

```
>>> x,*y,z  = 1,2,3,4,5
>>> x
1
```

```
>>> y
[2, 3, 4]
>>> z
5
```

This unpacking is sometimes called *de-structuring*, or *splat*, in case you read this term elsewhere.

Dictionaries Python dictionaries are central to Python because many other elements (e.g., functions, classes) are built around them. Effectively programming Python *means* using dictionaries effectively. Dictionaries are general containers that implement the *mapping* data structure, which is sometimes called a hash table or associative array. Dictionaries require a key/value pair, which maps the key to the value.

```
>>> x = {'key': 'value'}
```

The curly braces and the colon make the dictionary. To retrieve the value from the x dictionary, you must index it with the key as shown,

```
>>> x['key']
'value'
```

Let us get started with some basic syntax.

```
>>> x={'play':'Shakespeare','actor':'Wayne','direct':'Kubrick',
...      'author':'Hemmingway','bio':'Watson'}

>>> len(x) # number of key/value pairs
5
>>> x['pres']='F.D.R.' # assignment to key 'pres'
>>> x
{'play': 'Shakespeare', 'actor': 'Wayne', 'direct': 'Kubrick',
↪   'author': 'Hemmingway', 'bio': 'Watson', 'pres': 'F.D.R.'}
>>> x['bio']='Darwin' # reassignment for 'bio' key
>>> x
{'play': 'Shakespeare', 'actor': 'Wayne', 'direct': 'Kubrick',
↪   'author': 'Hemmingway', 'bio': 'Darwin', 'pres': 'F.D.R.'}
>>> del x['actor'] # delete key/value pair
>>> x
{'play': 'Shakespeare', 'direct': 'Kubrick', 'author':
↪   'Hemmingway', 'bio': 'Darwin', 'pres': 'F.D.R.'}
```

Dictionaries can also be created with the dict built-in function,

```
>>> # another way of creating a dict
>>> x=dict(key='value',
...         another_key=1.333,
...         more_keys=[1,3,4,'one'])
>>> x
{'key': 'value', 'another_key': 1.333, 'more_keys': [1, 3, 4,
↪   'one']}
>>> x={(1,3):'value'} # any immutable type can be a valid key
>>> x
{(1, 3): 'value'}
>>> x[(1,3)]='immutables can be keys'
```

As generalized containers, dictionaries can contain other dictionaries or lists or other Python types.

Programming Tip: Unions of Dictionaries
What if you want to create a union of dictionaries in one- line?

```
>>> d1 = {'a':1, 'b':2, 'c':3}
>>> d2 = {'A':1, 'B':2, 'C':3}
>>> dict(d1,**d2) # combo of d1 and d2
{'a': 1, 'b': 2, 'c': 3, 'A': 1, 'B': 2, 'C': 3}
>>> {**d1,**d2} # without dict function
{'a': 1, 'b': 2, 'c': 3, 'A': 1, 'B': 2, 'C': 3}
```

Pretty slick.

The invariant that the dictionary provides is that as long as you provide a valid key, then it will always retrieve the corresponding value; or, in the case of assignment, store the value reliably. Recall that lists are *ordered* data structures in the sense that when elements are indexed, the next element can be found by a relative offset from the prior one. This means that these elements are laid out contiguously in memory. Dictionaries do *not* have this property because they will put values wherever they can find memory, contiguous or not. This is because dictionaries do not rely upon relative offsets for indexing, they rely instead on a hash function. Consider the following,

```
>>> x = {0: 'zero', 1: 'one'}
>>> y = ['zero','one']
>>> x[1] # dictionary
'one'
>>> y[1] # list
'one'
```

Indexing both variables looks notationally the same in both cases, but the process is different. When given a key, the dictionary computes a hash function and the stores the value at a memory location based upon the hash function. What is a hash function? A hash function takes an input and is designed to return, with very high probability, a value that is unique to the key. In particular, this means that two keys cannot have the same hash, or, equivalently, cannot store different values in the same memory location. Here are two keys which are almost identical, but have very different hashes.

```
>>> hash('12345')
3973217705519425393
>>> hash('12346')
3824627720283660249
```

All this is with respect to probability, though. Because memory is finite, it could happen that the hash function produces values that are the same. This is known as a *hash collision* and Python implements fallback algorithms to handle this case. Nonetheless, as memory becomes scarce, especially on a small platform, the struggle to find suitable blocks of memory can be noticeable if your code uses many large dictionaries.

As we discussed before, inserting/removing elements from the middle of a list causes extra memory movement as the list maintains its invariant but this does *not* happen for dictionaries. This means that elements can be added or removed without any extra memory overhead beyond the cost of computing the hash function (i.e., constant-time lookup). Thus, dictionaries are ideal for codes that do *not* need ordering. Note that since Python 3.6+, dictionaries are *ordered* in the sense of the order in which items were inserted to the dictionary. In Python 2.7, this was known as `collections.OrderedDict` but has since become the default in Python 3.6+.

Now that we have a good idea of how dictionaries work, consider the inputs to the hash function: the keys. We have mainly used integers and strings for keys, but any immutable type can also be used, such as a tuple,

```
>>> x= {(1,3):10, (3,4,5,8):10}
```

However, if you try to use a mutable type as a key,

```
>>> a = [1,2]
>>> x[a]= 10
Traceback (most recent call last):
  File "<stdin>", line 1, in <module>
TypeError: unhashable type: 'list'
```

Let us think about why this happens. Remember that the hash function guarantees that when given a key it will always be able to retrieve the value. Suppose that it *were* possible use mutable keys in dictionaries. In the above code, we would have `hash(a) -> 132334`, as an example, and let us suppose the `10` value is inserted in that memory slot. Later in the code, we could change the contents of a as in `a[0]=3`. Now, because the hash function is guaranteed to produce different outputs for different inputs, the hash function output would be *different* from `132334` and thus the dictionary could not retrieve the corresponding value, which would violate its invariant. Thus we have arrived at a contradiction that explains why dictionary keys must be immutable.

Sets Python provides mathematical sets and corresponding operations with the `set()` data structure, which are basically dictionaries without values.

```
>>> set([1,2,11,1]) # union-izes elements
{1, 2, 11}
>>> set([1,2,3]) & set([2,3,4]) # bitwise intersection
{2, 3}
>>> set([1,2,3]) and set([2,3,4])
{2, 3, 4}
>>> set([1,2,3]) ^ set([2,3,4]) # bitwise exclusive OR
{1, 4}
>>> set([1,2,3]) | set([2,3,4]) # OR
{1, 2, 3, 4}
>>> set([ [1,2,3],[2,3,4] ]) # no sets of lists
```

```
(without more work)
Traceback (most recent call last):
  File "<stdin>", line 1, in <module>
TypeError: unhashable type: 'list'
```

Note that since Python 3.6+, keys can be used as set objects, as in the following,

```
>>> d = dict(one=1,two=2)
>>> {'one','two'} & d.keys() # intersection
{'one', 'two'}
>>> {'one','three'} | d.keys() # union
{'one', 'two', 'three'}
```

This also works for dictionary items if the values are hashable,

```
>>> d = dict(one='ball',two='play')
>>> {'ball','play'} | d.items()
{'ball', 'play', ('one', 'ball'), ('two', 'play')}
```

Once you create a set, you can add individual elements or remove them as follows:,

```
>>> s = {'one',1,3,'10'}
>>> s.add('11')
>>> s
{1, 3, 'one', '11', '10'}
>>> s.discard(3)
>>> s
{1, 'one', '11', '10'}
```

Remember sets are not ordered and you cannot directly index any of the constituent items. Also, the subset() method is for a *proper* subset not a partial subset. For example,

```
>>> a = {1,3,4,5}
>>> b = {1,3,4,5,6}
>>> a.issubset(b)
True
>>> a.add(1000)
>>> a.issubset(b)
False
```

And likewise for issuperset. Sets are optimal for fast lookups in Python, as in the following,

```
>>> a = {1,3,4,5,6}
>>> 1 in a
True
>>> 11 in a
False
```

which works *really* fast, even for large sets.

1.1.6   Loops and Conditionals

There are two primary looping constructs in Python: the for loop and the while loop. The syntax of the for loop is straightforward:

```
>>> for i in range(3):
...     print(i)
...
0
1
2
```

Note the colon character at the end. This is your hint that the next line should be indented. In Python, blocks are denoted by whitespace indentation (four spaces is recommended) which makes the code more readable anyway. The `for` loop iterates over items that provided by the `iterator`, which is the `range(3)` list in the above example. Python abstracts the idea of `iterable` out of the looping construction so that some Python objects are iterable all by themselves and are just waiting for an iteration provider like a `for` or `while` loop to get them going. Interestingly, Python has an `else` clause, which is used to determine whether or not the loop exited with a break[3] or not.

```
>>> for i in [1,2,3]:
...     if i>20:
...         break # won't happen
... else:
...     print('no break here!')
...
no break here!
```

The `else` block is only executed when the loop terminates without breaking.

The `while` loop has a similar straightforward construct:

```
>>> i = 0
>>> while i < 3:
...     i += 1
...     print(i)
...
1
2
3
```

This also has a corresponding optional `else` block. Again, note that the presence of the colon character hints at indenting the following line. The `while` loop will continue until the boolean expression (i.e., `i<10`) evaluates `False`. Let us consider boolean and membership in Python.

Logic and Membership Python is a *truthy* language in the sense that things are true except for the following:

* `None`
* `False`
* zero of any numeric type, for example, `0`, `0L`, `0.0`, `0j`.
* any empty sequence, for example, `"`, `()`, `[]`.

---

[3]There is also a `continue` statement which will jump to the top of the `for` or `while` loop.

- any empty mapping, for example, {}.
- instances of user-defined classes, if the class defines a __nonzero__() or __len__() method, when that method returns the integer zero or bool value False.

Let us try some examples,

```
>>> bool(1)
True
>>> bool([]) # empty list
False
>>> bool({}) # empty dictionary
False
>>> bool(0)
False
>>> bool([0,]) # list is not empty!
True
```

Python is syntactically clean about numeric intervals!

```
>>> 3.2 < 10 < 20
True
>>> True
True
```

You can use disjunctions (or), negations (not), and conjunctions (and) also.

```
>>> 1 < 2 and 2 < 3 or 3 < 1
True
>>> 1 < 2 and not 2 > 3  or 1<3
True
```

Use grouping parentheses for readability. You can use logic across iterables as in the following:

```
>>> (1,2,3) < (4,5,6) # at least one True
True
>>> (1,2,3) < (0,1,2) # all False
False
```

Do not use relative comparison for Python strings (i.e., 'a' < 'b') which is obtuse to read. Use string matching operations instead (i.e., ==). Membership testing uses the in keyword.

```
>>> 'on' in [22,['one','too','throw']]
False
>>> 'one' in [22,['one','too','throw']] # no recursion
False
>>> 'one' in [22,'one',['too','throw']]
True
>>> ['too','throw'] not in [22,'one',['too','throw']]
False
```

If you are testing membership across millions of elements it is *much* faster to use set() instead of list. For example,

```
>>> 'one' in {'one','two','three'}
True
```

The is keyword is stronger than equality, because it checks if two objects are the *same*.

```
>>> x = 'this string'
>>> y = 'this string'
>>> x is y
False
>>> x==y
True
```

However, is checks the id of each of the items:

```
>>> x=y='this string'
>>> id(x),id(y)
(140271930045360, 140271930045360)
>>> x is y
True
```

By virtue of this, the following idioms are common: x is True, x is None. Note that None is Python's singleton.

Conditionals Now that we understand boolean expressions, we can build conditional statements using if.

```
>>> if 1 < 2:
...       print('one less than two')
...
one less than two
```

There is an else and elif, but no switch statement.

```
>>> a = 10
>>> if a < 10:
...       print('a less than 10')
... elif a < 20:
...       print('a less than 20')
... else:
...       print('a otherwise')
...
a less than 20
```

There is a rare one-liner for conditionals

```
>>> x = 1 if (1>2) else 3 # 1-line conditional
>>> x
3
```

List Comprehensions Collecting items over a loop is so common that it is its own idiom in Python. That is,

```
>>> out=[] # initialize container
>>> for i in range(10):
...     out.append(i**2)
...
>>> out
[0, 1, 4, 9, 16, 25, 36, 49, 64, 81]
```

This can be abbreviated as a list comprehension.

```
>>> [i**2 for i in range(10)] # squares of integers
[0, 1, 4, 9, 16, 25, 36, 49, 64, 81]
```

Conditional elements can be embedded in the list comprehension.

```
>>> [i**2 for i in range(10) if i % 2] # embedded conditional
[1, 9, 25, 49, 81]
```

which is equivalent to the following,

```
>>> out = []
>>> for i in range(10):
...     if i % 2:
...         out.append(i**2)
...
>>> out
[1, 9, 25, 49, 81]
```

These comprehensions also work with dictionaries and sets.

```
>>> {i:i**2 for i in range(5)} # dictionary
{0: 0, 1: 1, 2: 4, 3: 9, 4: 16}
>>> {i**2 for i in range(5)}    # set
{0, 1, 4, 9, 16}
```

1.1.7 Functions

There are two common ways to define functions. You can use the `def` keyword as in the following:

```
>>> def foo():
...     return 'I said foo'
...
>>> foo()
'I said foo'
```

Note that you need a `return` statement and the trailing parenthesis to invoke the function. Without the `return` statement, the functions returns the `None` singleton. Functions are first-class objects.

```
>>> foo # just another Python object
<function foo at 0x7f939a6ecc10>
```

In the early days of Python, this was a key feature because otherwise only pointers to functions could be passed around and those required special handling. Practically speaking, firsts-class objects can be manipulated like any other Python object—they can be put in containers and passed around without any special handling. Naturally, we want to supply arguments to our functions. There are two kinds of function arguments *positional* and *keyword*.

```
>>> def foo(x): # positional argument
...        return x*2
...
>>> foo(10)
20
```

Positional arguments can be specified with their names,

```
>>> def foo(x,y):
...     print('x=',x,'y=',y)
...
```

The variables x and y can be specified by their position as follows:,

```
>>> foo(1,2)
x= 1 y= 2
```

These can also be specified using the names instead of the positions as follows:,

```
>>> foo(y=2,x=1)
x= 1 y= 2
```

Keyword arguments allow you to specify defaults.

```
>>> def foo(x=20): # keyword named argument
...        return 2*x
...
>>> foo(1)
2
>>> foo()
40
>>> foo(x=30)
60
```

You can have multiple specified defaults,

```
>>> def foo(x=20,y=30):
...        return x+y
...
>>> foo(20,)
50
>>> foo(1,1)
2
>>> foo(y=12)
32
>>> help(foo)
Help on function foo:

foo(x=20, y=30)
```

Programming Tip: Documenting Functions
Whenever possible, provide meaningful variable names and defaults for your functions along with documentation strings. This makes your code easy to navigate, understand, and use.

Python makes it easy to include documentation for your functions using *docstrings*. Thise makes the `help` function more useful for functions.

```
>>> def compute_this(position=20,velocity=30):
...       '''position in m
...       velocity in m/s
...       '''
...       return x+y
...
>>> help(compute_this)
Help on function compute_this:

compute_this(position=20, velocity=30)
    position in m
    velocity in m/s
```

Thus, by using meaningful argument and function names and including basic documentation in the docstrings, you can *greatly* improve the usability of your Python functions. Also, it is recommended to make function names verb-like (e.g., `get_this`, `compute_field`).

In addition to using the `def` statement, you can also create one-line functions using `lambda`. These are sometimes called *anonymous* functions.

```
>>> f = lambda x:   x**2 # anonymous functions
>>> f(10)
100
```

As first-class objects, functions can be put into lists just like any other Python object,

```
>>> [lambda x: x, lambda x:x**2] # list of functions
[<function <lambda> at 0x7f939a6ba700>, <function <lambda> at
↪    0x7f939a6ba790>]
>>> for i in   [lambda x: x, lambda x:x**2]:
...     print(i(10))
...
10
100
```

So far, we have not made good use of `tuple`, but these data structures become very powerful when used with functions. This is because they allow you to separate the function arguments from the function itself. This means you can pass them around and build function arguments and then later execute them with one or more functions. Consider the following function,

```
>>> def foo(x,y,z):
...       return x+y+z
...
>>> foo(1,2,3)
6
>>> args = (1,2,3)
>>> foo(*args) # splat tuple into arguments
6
```

The star notation in front of the tuple unpacks the tuple into the function signature. We have already seen this kind of unpacking assignment with tuples.

```
>>> x,y,z = args
>>> x
1
>>> y
2
>>> y
2
```

This is the same behavior for unpacking into the function signature. The double asterisk notation does the corresponding unpacking for keyword arguments,

```
>>> def foo(x=1,y=1,z=1):
...     return x+y+z
...
>>> kwds = dict(x=1,y=2,z=3)
>>> kwds
{'x': 1, 'y': 2, 'z': 3}
>>> foo(**kwds) # double asterisks for keyword splat
6
```

You can use both at the same time:

```
>>> def foo(x,y,w=10,z=1):
...     return (x,y,w,z)
...
>>> args = (1,2)
>>> kwds = dict(w=100,z=11)
>>> foo(*args,**kwds)
(1, 2, 100, 11)
```

Function Variable Scoping Variables within functions or subfunctions are local to that respective scope. Global variables require special handling if they are going to be changed inside the function body.

```
>>> x=10 # outside function
>>> def foo():
...     return x
...
>>> foo()
10
>>> print('x = %d is not changed'%x)
x = 10 is not changed

>>> def foo():
...     x=1 # defined inside function
...     return x
...
>>> foo()
1
>>> print('x = %d is not changed'%x)
x = 10 is not changed

>>> def foo():
...     global x # define as global
...     x=20     # assign inside function scope
...     return x
```

```
...
>>> foo()
20
>>> print('x = %d IS changed!'%x)
x = 20 IS changed!
```

Function Keyword Filtering Using **kwds at the end allows a function to disregard unused keywords while filtering out (using the function signature) the keyword inputs that it *does* use.

```
>>> def foo(x=1,y=2,z=3,**kwds):
...     print('in foo, kwds = %s'%(kwds))
...     return x+y+z
...
>>> def goo(x=10,**kwds):
...     print('in goo, kwds = %s'%(kwds))
...     return foo(x=2*x,**kwds)
...
>>> def moo(y=1,z=1,**kwds):
...     print('in moo, kwds = %s'%(kwds))
...     return goo(x=z+y,z=z+1,q=10,**kwds)
...
```

This means you can call any of these with an unspecified keyword as in

```
>>> moo(y=91,z=11,zeta_variable = 10)
in moo, kwds = {'zeta_variable': 10}
in goo, kwds = {'z': 12, 'q': 10, 'zeta_variable': 10}
in foo, kwds = {'q': 10, 'zeta_variable': 10}
218
```

and the zeta_variable will be passed through unused because no function uses it. Thus, you can inject some other function in the calling sequence that *does* use this variable, without having to change the call signatures of any of the other functions. Using keyword arguments this way is very common when using Python to wrap other codes.

Because this is such an awesome and useful Python feature, here is another example where we can trace how each of the function signatures is satisfied and the rest of the keyword arguments are passed through.

```
>>> def foo(x=1,y=2,**kwds):
...     print('foo: x = %d, y = %d, kwds=%r'%(x,y,kwds))
...     print('\t',)
...     goo(x=x,**kwds)
...
>>> def goo(x=10,**kwds):
...     print('goo: x = %d, kwds=%r'%(x,kwds))
...     print('\t\t',)
...     moo(x=x,**kwds)
...
>>> def moo(z=20,**kwds):
...     print('moo: z=%d, kwds=%r'%(z,kwds))
...
```

Then,

```
>>> foo(x=1,y=2,z=3,k=20)
foo: x = 1, y = 2, kwds={'z': 3, 'k': 20}

goo: x = 1, kwds={'z': 3, 'k': 20}

moo: z=3, kwds={'x': 1, 'k': 20}
```

Notice how the function signatures of each of the functions are satisfied and the rest of the keyword arguments are passed through.

Python 3 has the ability to force users to supply keywords arguments using the * symbol in the function signature,

```
>>> def foo(*,x,y,z):
...      return x*y*y
...
```

Then,

```
>>> foo(1,2,3)            # no required keyword arguments?
Traceback (most recent call last):
  File "<stdin>", line 1, in <module>
TypeError: foo() takes 0 positional arguments but 3 were given
>>> foo(x=1,y=2,z=3) # must use keywords
4
```

Using *args and **kwargs provides a general interface function arguments but these do *not* play nicely with integrated code development tools because the variable introspection does not work for these function signatures. For example,

```
>>> def foo(*args,**kwargs):
...      return args, kwargs
...
>>> foo(1,2,2,x=12,y=2,q='a')
((1, 2, 2), {'x': 12, 'y': 2, 'q': 'a'})
```

This leaves it up to the function to process the arguments, which makes for an unclear function signature. You should *avoid* this whenever possible.

Functional Programming Idioms   Although not a *real* functional programming language like Haskell, Python has useful functional idioms. These idioms become important in parallel computing frameworks like PySpark.

```
>>> map(lambda x:  x**2 , range(10))
<map object at 0x7f939a6a6fa0>
```

This applies the given (lambda) function to each of the iterables in range(10) but you have to convert it to a list to get the output.

```
>>> list(map(lambda x:  x**2 , range(10)))
[0, 1, 4, 9, 16, 25, 36, 49, 64, 81]
```

This gives the same output as the corresponding list comprehension,

```
>>> list(map(lambda x:  x**2, range(10)))
[0, 1, 4, 9, 16, 25, 36, 49, 64, 81]
>>> [i**2 for i in range(10)]
[0, 1, 4, 9, 16, 25, 36, 49, 64, 81]
```

There is also a `reduce` function,

```
>>> from functools import reduce
>>> reduce(lambda x,y:x+2*y,[0,1,2,3],0)
12
>>> [i for i in range(10) if i %2 ]
[1, 3, 5, 7, 9]
```

Pay attention to the recursive problem that `functools.reduce` solves because `functools.reduce` is super-fast in Python. For example, the least common multiple algorithm can be effectively implemented using `functools.reduce`, as shown:

```
>>> from functools import reduce
>>> def gcd(a, b):
...     'Return greatest common divisor using Euclids Algorithm.'
...     while b:
...         a, b = b, a % b
...     return a
...
>>> def lcm(a, b):
...     'Return lowest common multiple.'
...     return a * b // gcd(a, b)
...
>>> def lcmm(*args):
...     'Return lcm of args.'
...     return reduce(lcm, args)
...
```

Programming Tip: Beware Default Containers in Functions

```
>>> def foo(x=[]): # using empty list as default
...     x.append(10)
...     return x
...
>>> foo() # maybe you expected this...
[10]
>>> foo() # ... but did you expect this...
[10, 10]
>>> foo() # ... or this? What's going on here?
[10, 10, 10]
```

Unspecified arguments are usually handled with `None` in the function signature,

```
>>> def foo(x=None):
...     if x is None:
...         x = 10
...
```

The logic of the code resolves the missing items.

Function Deep Dive Let us see how Python constructs the function object with an example:

```
>>> def foo(x):
...     return x
...
```

The __code__ attribute of the foo function contains the internals of the function. For example, foo.__code__.co_argcount show the number of arguments for the foo function.

```
>>> foo.__code__.co_argcount
1
```

The co_varnames attribute gives the name of the argument as a tuple of strings,

```
>>> foo.__code__.co_varnames
('x',)
```

Local variables are also contained in the function object.

```
>>> def foo(x):
...     y= 2*x
...     return y
...
>>> foo.__code__.co_varnames
('x', 'y')
```

Recall that functions can also use *args for arbitrary inputs not specified at the time of function definition.

```
>>> def foo(x,*args):
...     return x+sum(args)
...
>>> foo.__code__.co_argcount # same as before?
1
```

The argument count did not increase because *args is handled with the co_flags attribute of the function object. This is a bitmask that indicates other aspects of the function object.

```
>>> print('{0:b}'.format(foo.__code__.co_flags))
1000111
```

Note that the third bit (i.e., $2\char`^2$ coefficient) is 1 which indicates that the function signature contains a *args entry. In hexadecimal, the 0x01 mask corresponds to co_optimized (use fast locals), 0x02 to co_newlocals (new dictionary for code block), 0x04 to co_varags (has *args), 0x08 to co_varkeywords (has **kwds in function signature), 0x10 to co_nested (nested function scope), and finally 0x20 to co_generator (function is a generator).

The dis module can help unpack the function object.

```
>>> def foo(x):
...     y= 2*x
...     return y
...
>>> import dis
>>> dis.show_code(foo)
Name:                foo
Filename:            <stdin>
Argument count:      1
Positional-only arguments: 0
Kw-only arguments: 0
Number of locals:  2
Stack size:        2
Flags:             OPTIMIZED, NEWLOCALS, NOFREE
Constants:
   0: None
   1: 2
Variable names:
   0: x
   1: y
```

Note that constants are *not* compiled into byte-code, but are stored in the function object and then referenced later in the byte-code. For example,

```
>>> def foo(x):
...     a,b = 1,2
...     return x*a*b
...
>>> print(foo.__code__.co_varnames)
('x', 'a', 'b')
>>> print(foo.__code__.co_consts)
(None, (1, 2))
```

where the None singleton is always available here for use in the function. Now, we can examine the byte-code in the co_code attribute in more detail using dis.dis

```
>>> print(foo.__code__.co_code) # raw byte-code
b'd\x01\\\x02}\x01}\x02|\x00|\x01\x14\x00|\x02\x14\x00S\x00'
>>> dis.dis(foo)
  2           0 LOAD_CONST            1 ((1, 2))
              2 UNPACK_SEQUENCE       2
              4 STORE_FAST            1 (a)
              6 STORE_FAST            2 (b)

  3           8 LOAD_FAST             0 (x)
             10 LOAD_FAST             1 (a)
             12 BINARY_MULTIPLY
             14 LOAD_FAST             2 (b)
             16 BINARY_MULTIPLY
             18 RETURN_VALUE
```

where LOAD_CONST takes the constants previously stored in the function and LOAD_FAST means use a local variable and the rest are self-explanatory. Functions store default values in the __defaults__ tuple. For example,

```
>>> def foo(x=10):
...     return x*10
```

```
...
>>> print(foo.__defaults__)
(10,)
```

The scoping rules for functions follow the order Local, Enclosing (i.e., closure), Global, Built-ins (LEGB). In the function body, when Python encounters a variable, it first checks to see if it is a local variable (co_varnames)) and then checks for the rest if it is not. Here is an interesting example,

```
>>> def foo():
...     print('x=',x)
...     x = 10
...
>>> foo()
Traceback (most recent call last):
  File "<stdin>", line 1, in <module>
  File "<stdin>", line 2, in foo
UnboundLocalError: local variable 'x' referenced before assignment
```

Why is this? Let us look at the function internals:

```
>>> foo.__code__.co_varnames
('x',)
```

When Python tries to resolve x, it checks to see if it is a local variable, and it is because it shows up in co_varnames, but there has been no assignment to it and therefore generates the UnboundLocalError.

Enclosing for function scoping is more involved. Consider the following code,

```
>>> def outer():
...     a,b = 0,0
...     def inner():
...         a += 1
...         b += 1
...         print(f'{a},{b}')
...     return inner
...
>>> f = outer()
>>> f()
Traceback (most recent call last):
  File "<stdin>", line 1, in <module>
  File "<stdin>", line 4, in inner
UnboundLocalError: local variable 'a' referenced before assignment
```

We saw this error above. Let us examine the function object,

```
>>> f.__code__.co_varnames
('a', 'b')
```

This means that the inner function thinks these variables are local to it when they actually exist in the enclosing function scope. We can fix this with the nonlocal keyword.

```
>>> def outer():
...     a,b = 0,0
```

```
...      def inner():
...          nonlocal a,b # use nonlocal
...          a+=1
...          b+=1
...          print(f'{a},{b}')
...      return inner
...
>>> f = outer()
>>> f() # this works now
1,1
```

If you go back and check the `co_varnames` attribute, you will see that it is empty.
The `co_freevars` in the inner function contains the info for the `nonlocal`
variables,

```
>>> f.__code__.co_freevars
('a', 'b')
```

so that the inner function knows what these are in the enclosing scope. The outer
function also knows which variables are being used in the embedded function via
the `co_cellvars` attribute.

```
>>> outer.__code__.co_cellvars
('a', 'b')
```

Thus this enclosing relationship goes both ways.

Function Stack Frames Nested Python functions are put on a stack. For example,

```
>>> def foo():
...      return 1
...
>>> def goo():
...      return 1+foo()
...
```

So when `goo` is called, it calls `foo` so that `foo` is on top of `goo` in the stack.
The *stack frame* is the data structure that maintains program scope and information
about the state of the execution. Thus, in this example, there are two frames for each
function with `foo` on the topmost level of the stack.

We can inspect the internal execution state via the stack frame using the
following,

```
>>> import sys
>>> depth = 0 # top of the stack
>>> frame = sys._getframe(depth)
>>> frame
<frame at 0x7f939a6bcba0, file '<stdin>', line 1, code <module>>
```

where `depth` is the number of calls below the top of the stack and 0 corresponds
to the current frame. Frames contain local variables (`frame.f_locals`) in the
current scope and global variables (`frame.f_globals`) in the current module. Note
that local and global variables can also be accessed using the `locals()` and
`globals()` built-in functions. The frame object also has the `f_lineno` (current

line number), `f_trace` (tracing function), and `f_back` (reference to prior frame) attributes. During an unhandled exception, Python navigates backwards to display the stack frames using `f_back`. It is important to delete the frame object otherwise there is the danger of creating a reference cycle.

The stack frame also contains `frame.f_code` which is executable byte-code. This is different from the function object because it does not contain the same references to the global execution environment. For example, a code object can be created using the `compile` built-in.

```
>>> c = compile('1 + a*b','tmp.py','eval')
>>> print(c)
<code object <module> at 0x7f939a6a5df0, file "tmp.py", line 1>
>>> eval(c,dict(a=1,b=2))
3
```

Note that `eval` evaluates the expression for the code object. The code object contains the `co_filename` (filename where created), `co_name` (name of function/module), `co_varnames` (names of variables), and `co_code` (compiled byte-code) attributes.

Programming Tip: Asserts in Functions

Writing clean and reusable functions is fundamental to effective Python programming. The easiest way to vastly improve your functions' reliability and reusability is sprinkle `assert` statements into your code. These statements raise an `AssertionError` if `False` and provide a good way to ensure that your function is behaving as expected. Consider the following example,

```
>>> def foo(x):
...     return 2*x
...

>>> foo(10)
20
>>> foo('x')
'xx'
```

If the intention of the function is work with numerical inputs, then the `foo('x')` slipped right through, but is still valid Python. This is called *fail-through* and is the most insidious byproduct of dynamic typing. A quick way to fix this is to `assert` the input as in the following,

```
>>> def foo(x):
...     assert isinstance(x,int)
...     return 2*x
...
>>> foo('x')
Traceback (most recent call last):
  File "<stdin>", line 1, in <module>
```

(continued)

```
    File "<stdin>", line 2, in foo
AssertionError
```

Now, the function is restricted to work with integer inputs and will raise
`AssertionError` otherwise. Beyond checking types, `assert` statements
can ensure the business logic of your functions by checking that intermediate
computations are always positive, add up to one, or whatever. Also, there
is a command line switch on `python` that will allow you to turn off
`assert` statements, if need be; but your `assert` statements should not
be excessively complicated, but rather should provide a fallback position
for unexpected usages. Think of `assert` statements and pre-placement of
debugger breakpoints that you can use later.

Clear and expressive functions are the hallmark of solid Python programming.
Using variable and function names that are clear, with corresponding detailed
docstrings means that users of your functions (including later versions of yourself)
will thank you. A good general rule of thumb is that your docstring should
be *longer* than the body of your function. If not, then break up your function
into smaller chunks. For examples of outstanding Python code, check out the
networkx[3] project.

Lazy Evaluation and Function Signatures Consider the following function,

```
>>> def foo(x):
...     return a*x
...
```

Note that the interpreter does not complain when you define this function even
though the variable a is not defined. It will only complain when you try to run the
function. This is because Python functions are lazy evaluated, which means that the
function bodies are only processed when called. Thus, the function does not know
about the missing a variable until it tries to look it up in the namespace. Although
the function bodies are lazy evaluated, the function signature is eagerly evaluated.
Consider the following experiment,

```
>>> def report():
...     print('I was called!')
...     return 10
...
>>> def foo(x = report()):
...     return x
...
I was called!
```

Note that the foo was not called but `report()` was because it appears in the
function signature of foo.

---

[3] See https://networkx.org/.

1.1.8   File Input/Output

It is straightforward to read and write files using Python. The same pattern applies
when writing to other objects, like sockets. The following is the traditional way to
get file I/O in Python.

```
>>> f=open('myfile.txt','w') # write mode
>>> f.write('this is line 1')
14
>>> f.close()
>>> f=open('myfile.txt','a') # append mode
>>> f.write('this is line 2')
14
>>> f.close()
>>> f=open('myfile.txt','a') # append mode
>>> f.write('\nthis is line 3\n')  # put in newlines
16
>>> f.close()
>>> f=open('myfile.txt','a') # append mode
>>> f.writelines([ 'this is line 4\n', 'this is line 5\n']) #
↪    put in newlines
>>> f=open('myfile.txt','r') # read mode
>>> print(f.readlines())
['this is line 1this is line 2\n', 'this is line 3\n', 'this is
↪    line 4\n', 'this is line 5\n']
>>> ['this is line 1this is line 2\n', 'this is line 3\n', 'this
↪    is line 4\n', 'this is line 5\n']
['this is line 1this is line 2\n', 'this is line 3\n', 'this is
↪    line 4\n', 'this is line 5\n']
```

Once you have the file handle, you can use methods like seek() to move the
pointer around in the file. Instead of having to explicitly close the file, the with
statement handles this automatically,

```
>>> with open('myfile.txt','r') as f:
...     print(f.readlines())
...
['this is line 1this is line 2\n', 'this is line 3\n', 'this is
↪    line 4\n', 'this is line 5\n']
```

The main advantage is that the file will automatically be closed at the end of the
block, which means you do not have to remember to close it later using f.close.
Briefly, when the with block is entered, Python runs the f.__enter__ method
to open the file and then at the end of the block runs the f.__exit__ method. The
with statement works with other conformable objects that respect this protocol and
with context managers.[4]

Note that for writing non-text files, you should use the rb read-binary and wb
write-binary equivalents of the above. You can also specify a file encoding with the
open function.

---

[4]See the contextlib built-in module.

> **Programming Tip: Other I/O Modules**
> Python has many tools for handling file I/O at different levels of granularity.
> The `struct` module is good for pure binary reading and writing. The `mmap`
> module is useful for bypassing the filesystem and using virtual memory for
> fast file access. The StringIO module allows read and write of strings as files.

Serialization: Saving Complex objects Serialization means packing Python
objects to be shipped between separate Python processes or separate computers,
say, through a network socket. The multiplatform nature of Python means that
one cannot be assured that the low-level attributes of Python objects (say, between
platform types or Python versions) will remain consistent.

For the vast majority of situations, the following will work.

```
>>> import pickle
>>> mylist = ["This", "is", 4, 13327]
>>> f=open('myfile.dat','wb') # binary-mode
>>> pickle.dump(mylist, f)
>>> f.close()
>>> f=open('myfile.dat','rb') # write mode
>>> print(pickle.load(f))
['This', 'is', 4, 13327]
```

> **Programming Tip: Serialization via Sockets**
> You can also serialize strings and transport them with intermediate file
> creation, using sockets, or some other protocol.

Pickling a Function The internal state of a function and how it is hooked into
the Python process in which it was created makes it tricky to pickle functions. As
with everything in Python, there are ways around this, depending on your use-
case. One idea is to use the `marshal` module to dump the function object in
a binary format, write it to a file, and then reconstruct it on the other end using
`types.FunctionType`. The downside of this technique is that it may not be
compatible across different major Python versions, even if they are all CPython
implementations.

```
>>> import marshal
>>> def foo(x):
...     return x*x
...
>>> code_string = marshal.dumps(foo.__code__)
```

Then in the remote process (after transferring `code_string`):

```
>>> import marshal, types
>>> code = marshal.loads(code_string)
```

```
>>> func = types.FunctionType(code, globals(), "some_func_name")
>>> func(10)  # gives 100
100
```

> **Programming Tip: Dill Pickles**
> The dill module can pickle functions and handle all these complications.
> However, import dill hijacks *all* pickling from that point forward. For
> more fine-grained control of serialization using dill without this hijacking,
> do dill.extend(False) after importing dill.
>
> ```
> import dill
> def foo(x):
>     return x*x
>
> dill.dumps(foo)
> ```

1.1.9   Dealing with Errors

This is where Python really shines. The approach is to ask for forgiveness rather
than permission. This is the basic template.

```
try:
  # try something
except:
  # fix something
```

The above except block will capture and process any kind of exception that is
thrown in the try block. Python provides a long list of built-in exceptions that you
can catch and the ability to create your own exceptions, if needed. In addition to
catching exceptions, you can raise your own exception using the raise statement.
There is also an assert statement that can throw exceptions if certain statements
are not True upon assertion (more on this later).

The following are some examples of the exception handling powers of Python.

```
>>> def some_function():
...     try:
...         # Division by zero raises an exception
...         10 / 0
...     except ZeroDivisionError:
...         print("Oops, invalid.")
...     else:
...         # Exception didn't occur, we're good.
...         pass
...     finally:
...         # This is executed after the code block is run
...         # and all exceptions have been handled, even
...         # if a new exception is raised while handling.
```

```
...            print("We're done with that.")
...
>>> some_function()
Oops, invalid.
We're done with that.
>>> out = list(range(3))
```

Exceptions can be very specific,

```
>>> try:
...     10 / 0
... except ZeroDivisionError:
...     print('I caught an attempt to divide by zero')
...
I caught an attempt to divide by zero
>>> try:
...     out[999]   # raises IndexError
... except ZeroDivisionError:
...     print('I caught an attempt to divide by zero')
...
Traceback (most recent call last):
  File "<stdin>", line 2, in <module>
IndexError: list index out of range
```

The exceptions that are caught change the code flow,

```
>>> try:
...     1/0
...     out[999]
... except ZeroDivisionError:
...     print('I caught an attempt to divide by zero but I did not
↪   try out[999]')
...
I caught an attempt to divide by zero but I did not try out[999]
```

The order of the exceptions in the try block matters,

```
>>> try:
...     1/0         # raises ZeroDivisionError
...     out[999]   # never gets this far
... except IndexError:
...     print('I caught an attempt to index something out of
↪   range')
...
Traceback (most recent call last):
  File "<stdin>", line 2, in <module>
ZeroDivisionError: division by zero
```

Blocks can be nested but if these get more than two layers deep, it bodes poorly for the code overall.

```
>>> try: #nested exceptions
...    try: # inner scope
...        1/0
...    except IndexError:
...        print('caught index error inside')
... except ZeroDivisionError as e:
```

```
...     print('I caught an attempt to divide by zero inside
↪   somewhere')
...
I caught an attempt to divide by zero inside somewhere
```

The `finally` clause always runs.

```
>>> try:   #nested exceptions with finally clause
...     try:
...         1/0
...     except IndexError:
...         print('caught index error inside')
...     finally:
...         print("I am working in inner scope")
... except ZeroDivisionError as e:
...     print('I caught an attempt to divide by zero inside
↪   somewhere')
...
I am working in inner scope
I caught an attempt to divide by zero inside somewhere
```

Exceptions can be grouped in tuples,

```
>>> try:
...     1/0
... except (IndexError,ZeroDivisionError) as e:
...     if isinstance(e,ZeroDivisionError):
...         print('I caught an attempt to divide by zero inside
↪   somewhere')
...     elif isinstance(e,IndexError):
...         print('I caught an attempt to index something out of
↪   range')
...
I caught an attempt to divide by zero inside somewhere
```

Although you can catch *any* exception with an unqualified `except` line, that would not tell you what exception was thrown, but you can use `Exception` to reveal that.

```
>>> try: # more detailed arbitrary exception catching
...     1/0
... except Exception as e:
...     print(type(e))
...
<class 'ZeroDivisionError'>
```

Programming Tip: Using Exceptions to Control Function Recursion
As discussed previously, nested function calls result in more stack frames which eventually hits a recursion limit. For example, the following recursive function will eventually fail with large enough n.

(continued)

```
>>> def factorial(n):
...     if (n == 0):   return 1
...     return n * factorial(n-1)
...
```

Python exceptions can be used to stop additional frames piling on the stack. The following `Recurse` is a subclass of `Exception` and does little more than save the arguments. The `recurse` function importantly raises the `Recurse` exception which stops growing the stack when called. The purpose of these two definitions is to create the `tail_recursive` decorator within which the real work occurs. The decorator returns a function that embeds an infinite while-loop that can exit only when the so-decorated function terminates without triggering an additional recursive step. Otherwise, the input arguments to the decorated function are passed on and the input arguments actually hold the intermediate values of the recursive calculation. This technique bypasses the built-in Python recursion limit and will work as a decorator for *any* recursive function that uses input arguments to store intermediate values.[a]

```
>>> class Recurse(Exception):
...     def __init__(self, *args, **kwargs):
...         self.args, self.kwargs = args, kwargs
...
>>> def recurse(*args, **kwargs):
...     raise Recurse(*args, **kwargs)
...
>>> def tail_recursive(f):
...     def decorated(*args, **kwargs):
...         while True:
...             try:
...                 return f(*args, **kwargs)
...             except Recurse as r:
...                 args, kwargs = r.args, r.kwargs
...     return decorated
...
>>> @tail_recursive
... def factorial(n, accumulator=1):
...     if n == 0: return accumulator
...     recurse(n-1, accumulator=accumulator*n)
...
```

---

[a] See https://chrispenner.ca/posts/python-tail-recursion for in-depth discussion of this technique.

1.1.10 Power Python Features to Master

Using these Python built-ins indicates maturity in your Python coding skills.

The `zip` Function Python has a built-in `zip` function that can combine iterables pair-wise.

```
>>> zip(range(3),'abc')
<zip object at 0x7f939a91d200>

>>> list(zip(range(3),'abc'))
[(0, 'a'), (1, 'b'), (2, 'c')]
>>> list(zip(range(3),'abc',range(1,4)))
[(0, 'a', 1), (1, 'b', 2), (2, 'c', 3)]
```

The slick part is reversing this operation using the `*` operation,

```
>>> x = zip(range(3),'abc')
>>> i,j = list(zip(*x))
>>> i
(0, 1, 2)
>>> j
('a', 'b', 'c')
```

When combined with `dict`, `zip` provides a powerful way to build Python dictionaries,

```
>>> k = range(5)
>>> v = range(5,10)
>>> dict(zip(k,v))
{0: 5, 1: 6, 2: 7, 3: 8, 4: 9}
```

The `max` Function The `max` function takes the maximum of a sequence.

```
>>> max([1,3,4])
4
```

If the items in the sequence are tuples, then the *first* item in the tuple is used for the ranking

```
>>> max([(1,2),(3,4)])
(3, 4)
```

This function takes a `key` argument that controls how the items in the sequence are evaluated. For example, we can rank based on the second element in the tuple,

```
>>> max([(1,4),(3,2)], key=lambda i:i[1])
(1, 4)
```

The `key` argument also works with the `min` and `sorted` functions.

The `with` Statement Set up a context for subsequent code

```
>>> class ControlledExecution:
...     def __enter__(self):
...         #set things up
```

```
...           print('I am setting things up for you!')
...           return 'something to use in the with-block'
...       def __exit__(self, type, value, traceback):
...           #tear things down
...           print('I am tearing things down for you!')
...
>>> with ControlledExecution() as thing:
...       # some code
...       pass
...
I am setting things up for you!
I am tearing things down for you!
```

Most commonly, `with` statements are used with files:

```
f = open("sample1.txt") # file handle
f.__enter__()
f.read(1)
f.__exit__(None, None, None)
f.read(1)
```

This is the better way to open close files that makes it harder to forget to close the files when you are done.

```
with open("x.txt") as f:
    data = f.read()
    #do something with data
```

contextlib for Fast Context Construction The `contextlib` module makes it very easy to quickly create context managers.

```
>>> import contextlib
>>> @contextlib.contextmanager
... def my_context():
...     print('setting up ')
...     try:
...         yield {} # can yield object if necessary for 'as' part
...     except:
...         print('catch some errors here')
...     finally:
...         print('tearing down')
...
>>> with my_context():
...     print('I am in the context')
...
setting up
I am in the context
tearing down
>>> with my_context():
...     raise RuntimeError ('I am an error')
...
setting up
catch some errors here
tearing down
```

Python Memory Management Earlier, we used the `id` function to get the unique identifier of each Python object. In the CPython implementation, this is actually the memory location of the object. Any object in Python has a *reference counter* that keeps track of the labels that point to it. We saw this earlier in our discussion of lists with different labels (i.e., variable names). When there are no more labels pointing to a given object, then the memory for that object is released. This works fine except for container objects that can generate cyclical references. For example,

```
>>> x = [1,2]
>>> x.append(x) # cyclic reference
>>> x
[1, 2, [...]]
```

In this situation, the reference counter will never count down to zero and be released. To defend this, Python implements a garbage collector, which periodically stops the main thread of execution in order to find and remove any such references. The following code uses the powerful `ctypes` module to get direct access to the reference counter field in the C-structure in `object.h` for CPython.

```
>>> def get_refs(obj_id):
...        from ctypes import Structure, c_long
...        class PyObject(Structure):
...            _fields_ = [("reference_cnt", c_long)]
...        obj = PyObject.from_address(obj_id)
...        return obj.reference_cnt
...
```

Let us return to our example and see the number of references that point to the list labeled x.

```
>>> x = [1,2]
>>> idx = id(x)
>>> get_refs(idx)
1
```

Now, after we create the circular reference,

```
>>> x.append(x)
>>> get_refs(idx)
2
```

Even deleting x does not help,

```
>>> del x
>>> get_refs(idx)
1
```

We can force the garbage collector to work with the following,

```
>>> import gc
>>> gc.collect()
216
>>> get_refs(idx) # finally removed!
0
```

So we can finally get rid of that x list. Importantly, you should *not* have to manually fire the garbage collector because Python decides when to do that efficiently. A key aspect of these algorithms is whether or not the container is mutable. Thus, from a garbage collection standpoint, it is better to use tuples instead of lists because tuples do not have to be policed by the garbage collector like regular lists do.

1.1.11 Generators

Generators provide just-in-time memory-efficient containers.

- Produces a stream of on-demand values
- Only executes on next()
- The yield() function produces a value, but saves the function's state for later
- Usable exactly once (i.e., not re-usable after used the first time)

```
>>> def generate_ints(N):
...     for i in range(N):
...         yield i # the yield makes the function a generator
...
>>> x=generate_ints(3)
>>> next(x)
0
>>> next(x)
1
>>> next(x)
2
>>> next(x)
Traceback (most recent call last):
  File "<stdin>", line 1, in <module>
StopIteration
```

Emptying the generator raises the StopIteration exception. What is happening in the next block?

```
>>> next(generate_ints(3))
0
>>> next(generate_ints(3))
0
>>> next(generate_ints(3))
0
```

The problem is that the generator is not saved to a variable wherein the current state of the generator can be stored. Thus, the above code creates a *new* generator at every line, which starts the iteration at the beginning for each line. You can also iterate on these directly:

```
>>> for i in generate_ints(5): # no assignment necessary here
...     print(i)
...
0
```

```
1
2
3
4
```

Generators maintain an internal state that can be returned to after the `yield`. This means that generators can pick up where they left off.

```
>>> def foo():
...     print('hello')
...     yield 1
...     print('world')
...     yield 2
...
>>> x = foo()
>>> next(x)
hello
1

>>> # do some other stuff here
>>> next(x) # pick up where I left off
world
2
```

Generators can implement algorithms that have infinite loops.

```
>>> def pi_series(): # infinite series converges to pi
...     sm = 0
...     i = 1.0; j = 1
...     while True: # loops forever!
...         sm = sm + j/i
...         yield 4*sm
...         i = i + 2; j = j * -1
...
>>> x = pi_series()
>>> next(x)
4.0
>>> next(x)
2.666666666666667
>>> next(x)
3.466666666666667
>>> next(x)
2.8952380952380956
>>> next(x)
3.3396825396825403
>>> next(x)
2.9760461760461765
>>> next(x)
3.2837384837384844
>>> gen = (i for i in range(3))   # List comprehension style
```

You can also `send()` to an existing generator by putting the `yield` on the right side of the equal sign,

```
>>> def foo():
...     while True:
...         line=(yield)
```

```
...          print(line)
...
>>> x= foo()
>>> next(x) # get it going
>>> x.send('I sent this to you')
I sent this to you
```

These can be daisy-chained also:

```
>>> def goo(target):
...     while True:
...         line=(yield)
...         target.send(line.upper()+'---')
...
>>> x= foo()
>>> y= goo(x)
>>> next(x) # get it going
>>> next(y) # get it going
>>> y.send('from goo to you')
FROM GOO TO YOU---
```

Generators can also be created as list comprehensions by changing the bracket notation to parentheses.

```
>>> x= (i for i in range(10))
>>> print(type(x))
<class 'generator'>
```

Now, x is a generator. The `itertools` module is key to using generators effectively. For example, we can clone generators,

```
>>> x = (i for i in range(10))
>>> import itertools as it
>>> y, = it.tee(x,1) # clone generator
>>> next(y) # step this one
0
>>> list(zip(x,y))
[(1, 2), (3, 4), (5, 6), (7, 8)]
```

This delayed execution method becomes particularly useful when working with large data sets.

```
>>> x = (i for i in range(10))
>>> y = map(lambda i:i**2,x)
>>> y
<map object at 0x7f939af36fd0>
```

Note that y is also a generator and that nothing has been computed yet. You can also map functions onto sequences using `it.starmap`.

Yielding from Generators   The following idiom is common to iterate over a generator,

```
>>> def foo(x):
...     for i in x:
...         yield i
...
```

Then you can feed the generator x into foo,

```
>>> x = (i**2 for i in range(3)) # create generator
>>> list(foo(x))
[0, 1, 4]
```

With yield from, this can be done in one line

```
>>> def foo(x):
...     yield from x
...
```

Then,

```
>>> x = (i**2 for i in range(3)) # recreate generator
>>> list(foo(x))
[0, 1, 4]
```

There is much more that yield from can do, however. Suppose we have a generator/coroutine that receives data.

```
>>> def accumulate():
...     sm = 0
...     while True:
...         n = yield # receive from send
...         print(f'I got {n} in accumulate')
...         sm+=n
...
```

Let us see how this works,

```
>>> x = accumulate()
>>> x.send(None) # kickstart coroutine
>>> x.send(1)
I got 1 in accumulate
>>> x.send(2)
I got 2 in accumulate
```

What if you have a composition of functions and you wanted to pass sent values down into the embedded coroutine?

```
>>> def wrapper(coroutine):
...     coroutine.send(None) # kickstart
...     while True:
...         try:
...             x = yield           # capture what is sent...
...             coroutine.send(x) # ... and pass it thru
...         except StopIteration:
...             pass
...
```

Then, we could do something like this,

```
>>> w = wrapper(accumulate())
>>> w.send(None)
>>> w.send(1)
I got 1 in accumulate
>>> w.send(2)
I got 2 in accumulate
```

Note how the sent values pass directly through to the embedded coroutine. For fun, we can wrap this twice,

```
>>> w = wrapper(wrapper(accumulate()))
>>> w.send(None)
>>> w.send(1)
I got 1 in accumulate
>>> w.send(2)
I got 2 in accumulate
```

Now that we know how that works, the `wrapper` can be abbreviated as the following

```
>>> def wrapper(coroutine):
...     yield from coroutine
...
```

and everything would work the same (try it!). Additionally, this automatically handles embedded errors with the same transparency. A useful example is the case where you want to flatten a list of embedded containers like

```
>>> x = [1,[1,2],[1,[3]]]
>>> def flatten(seq):
...     for item in seq:
...         if hasattr(item,'__iter__'):
...             yield from flatten(item)
...         else:
...             yield item
...
>>> list(flatten(x))
[1, 1, 2, 1, 3]
```

There is another syntax that is used for generators that makes it possible to simultaneously send/receive. One obvious problem with our previous `accumulate` function is that you do not receive the accumulated value. That is remedied by changing one line of the previous code,

```
>>> def accumulate():
...     sm = 0
...     while True:
...         n = yield sm   # receive from send and emit sm
...         print (f'I got {n} in accumulate and sm ={sm}')
...         sm+=n
...
```

Then, we can do,

```
>>> x = accumulate()
>>> x.send(None) # still need to kickstart
0
```

Note that it returned zero.

```
>>> x.send(1)
I got 1 in accumulate and sm =0
1
```

```
>>> x.send(2)
I got 2 in accumulate and sm =1
3
>>> x.send(3)
I got 3 in accumulate and sm =3
6
```

1.1.12   Decorators

Decorators are functions that make functions out of functions. It sounds redundant, but turns out to be very useful for combining disparate concepts. In the code below, notice that the input to my_decorator is a function. Because we do not know the input arguments for that function, we have to pass them through into the new_function that is defined in the body using args and kwargs. Then, within the new_function, we explicitly call the input function fn and pass in those arguments. The most important line is the last line where we return the function defined within the body of my_decorator. Because that function includes the explicit call to fn, it replicates the functionality of fn, but we are free to do other tasks within the body of new_function.

```
>>> def my_decorator(fn): # note that function as input
...     def new_function(*args,**kwargs):
...         print('this runs before function')
...         return fn(*args,**kwargs) # return a function
...     return new_function
...
>>> def foo(x):
...     return 2*x
...
>>> goo = my_decorator(foo)
>>> foo(3)
6
>>> goo(3)
this runs before function
6
```

In the output above, notice that goo faithfully reproduces goo but with the extra output we put into the body of new_function. The important idea is that whatever the new functionality we built into the decorator, it should be *orthogonal* to the business logic of the input function. Otherwise, the identity of the input function gets mixed into the decorator, which becomes hard to debug and understand later. The following log_arguments decorator is a good example of this principle. Suppose we want to monitor a function's input arguments. The log_arguments decorator additionally prints out the input arguments to the input function but does *not* interfere with the business logic of that underlying function.

```
>>> def log_arguments(fn): # note that function as input
...     def new_function(*args,**kwargs):
...         print('positional arguments:')
```

```
...          print(args)
...          print('keyword arguments:')
...          print(kwargs)
...          return fn(*args,**kwargs) # return a function
...     return new_function
...
```

You can stack a decorator on top of a function definition using the @ syntax. The advantage is that you can keep the original function name, which means that downstream users do not have to keep track of another decorated version of the function.

```
>>> @log_arguments # these are stackable also
... def foo(x,y=20):
...     return x*y
...
>>> foo(1,y=3)
positional arguments:
(1,)
keyword arguments:
{'y': 3}
3
```

Decorators are very useful for caches, which avoid re-computing expensive function values,

```
>>> def simple_cache(fn):
...     cache = {}
...     def new_fn(n):
...         if n in cache:
...             print('FOUND IN CACHE; RETURNING')
...             return cache[n]
...         # otherwise, call function
...         # & record value
...         val = fn(n)
...         cache[n] = val
...         return val
...     return new_fn
...

>>> def foo(x):
...     return 2*x
...
>>> goo = simple_cache(foo)
>>> [goo(i) for i in range(5)]
[0, 2, 4, 6, 8]
>>> [goo(i) for i in range(8)]
FOUND IN CACHE; RETURNING
FOUND IN CACHE; RETURNING
FOUND IN CACHE; RETURNING
FOUND IN CACHE; RETURNING
FOUND IN CACHE; RETURNING
[0, 2, 4, 6, 8, 10, 12, 14]
```

The simple_cache decorator runs the input function but then stores each output in the cache dictionary with keys corresponding to function inputs. Then, if the function is called again with the same input, the corresponding function value is *not* re-computed, but is instead retrieved from cache, which can be very efficient if the input function takes a long time to compute. This pattern is so common it is now in functools.lru_cache in the Python standard library.

> **Programming Tip: Modules of Decorators**
> Some Python modules are distributed as decorators (e.g., the click module for creating commandline interfaces) to make it easy to bolt-on new functionality without changing the source code. The idea is whatever the new functionality that the decorator provides, it should be distinct from the business logic of the function that is being decorated. This *separates the concerns* between the decorator and the decorated function.

Decorators are also useful for executing certain functions in threads. Recall that a thread is a set of instructions that the CPU can run separately from the parent process. The following decorator wraps a function to run in a separate thread.

```
>>> def run_async(func):
...     from threading import Thread
...     from functools import wraps
...     @wraps(func)
...     def async_func(*args, **kwargs):
...         func_hl = Thread(target = func,
...                          args = args,
...                          kwargs = kwargs)
...         func_hl.start()
...         return func_hl
...     return async_func
...
```

The wrap function from the functools module fixes the function signature. This decorator is useful when you have a small side-job (like a notification) that you want to run out of the main thread of execution. Let us write a simple function that does some fake *work*.

```
>>> from time import sleep
>>> def sleepy(n=1,id=1):
...     print('item %d sleeping for %d seconds...'%(id,n))
...     sleep(n)
...     print('item %d done sleeping for %d seconds'%(id,n))
...
```

Consider the following code block:

```
>>> sleepy(1,1)
item 1 sleeping for 1 seconds...
item 1 done sleeping for 1 seconds
>>> print('I am here!')
I am here!
>>> sleepy(2,2)
```

```
item 2 sleeping for 2 seconds...
item 2 done sleeping for 2 seconds
>>> print('I am now here!')
I am now here!
```

Using the decorator, we can make asynchronous versions of this function:

```
@run_async
def sleepy(n=1,id=1):
    print('item %d sleeping for %d seconds...'%(id,n))
    sleep(n)
    print('item %d done sleeping for %d seconds'%(id,n))
```

And with the same block of statements above, we obtain the following sequence of printed outputs:

```
sleepy(1,1)
print('I am here!')
sleepy(2,2)
print('I am now here!')

I am here!
item 1 sleeping for 1 seconds...
item 2 sleeping for 2 seconds...
I am now here!
item 1 done sleeping for 1 seconds
item 2 done sleeping for 2 seconds
```

Notice that the last print statement in the block actually executed *before* the individual functions were completed. That is because the main thread of execution handles those print statements while the separate threads are sleeping for different amounts of time. In other words, in the first example the last statement is *blocked* by the previous statements and has to wait for them to finish before it can do the final print. In the second case, there is no blocking so it can get to the last statement right away while the other work goes on in separate threads.

Another common use of decorators is to create *closures*. For example,

```
>>> def foo(a=1):
...     def goo(x):
...         return a*x # uses `a` from outer scope
...     return goo
...
>>> foo(1)(10) # version with a=1
10
>>> foo(2)(10) # version with a=2
20
>>> foo(3)(10) # version with a=2
30
```

In this case, note that the embedded goo function requires a parameter a that is not defined in its function signature. Thus, the foo function manufactures different versions of the embedded goo function as shown. For example, suppose you have many users with different certificates to access data that are accessible via goo. Then foo can close the certificates over goo so that every user effectively has her own version of the goo function with the corresponding embedded certificate. This simplifies the code because the business logic and function signature of goo need not change and the certificate will automatically disappear when goo goes out of scope.

Programming Tip: Python Threads
Threads in Python are primarily used for making applications responsive. For example, if you have a GUI with lots of buttons and you want the application to react to each button as it is clicked, even though the application itself is busy rendering the display, then threads are the right tool. As another example, suppose you are downloading files from multiple websites, then doing each website using a separate thread makes sense because the availability of the content on each of the sites will vary. This is something you would not know ahead of time.

The main difference between threads and processes is that processes have their own compartmentalized resources. The C-language Python (i.e., CPython) implements a Global Interpreter Lock (GIL) that prevents threads from fighting over internal data structures. Thus, the GIL employs a course-grained locking mechanism where only one thread has access to resources at any time. The GIL thus simplifies thread programming because running multiple threads simultaneously requires complicated bookkeeping. The downside of the GIL is that you cannot run multiple threads simultaneously to speed up compute-constrained tasks. Note that certain alternative implementations of Python like IronPython use a finer-grain threading design rather than the GIL approach.

As a final comment, on modern systems with multiple cores, it could be that multiple threads actually slow things down because the operating system may have to switch threads between different cores. This creates additional overheads in the thread switching mechanism that ultimately slows things down. CPython implements the GIL at the level of byte-code, which means the byte-code instructions across different threads are prohibited from simultaneously execution.

1.1.13   Iteration and Iterables

Iterators permit finer control for looping constructs.

```
>>> a = range(3)
>>> hasattr(a,'__iter__')
True
>>> # generally speaking, returns object that supports iteration
>>> iter(a)
```

```
<range_iterator object at 0x7f939a6a6630>
>>> hasattr(_,'__iter__')
True
>>> for i in a: #use iterables in loops
...     print(i)
...
0
1
2
```

The `hasattr` check above is the traditional way to check if a given object is iterable. Here is the modern way,

```
>>> from collections.abc import Iterable
>>> isinstance(a,Iterable)
True
```

You can also use `iter()` with functions to create sentinels,

```
>>> x=1
>>> def sentinel():
...     global x
...     x+=1
...     return x
...
>>> for k in iter(sentinel,5): # stops when x = 5
...     print(k)
...
2
3
4
>>> x
5
```

You can use this with a file object `f` as in `iter(f.readline,")` which will read lines from the file until the end.

Enumeration   The standard library has an `enum` module to support enumeration. Enumeration means binding symbolic names to unique, constant values. For example,

```
>>> from enum import Enum
>>> class Codes(Enum):
...     START = 1
...     STOP = 2
...     ERROR = 3
...
>>> Codes.START
<Codes.START: 1>
>>> Codes.ERROR
<Codes.ERROR: 3>
```

Once these have been defined, trying to change them will raise an `AttributeError`. Names and values can be extracted,

```
>>> Codes.START.name
'START'
>>> Codes.START.value
1
```

Enumerations can be iterated over,

```
>>> [i for i in Codes]
[<Codes.START: 1>, <Codes.STOP: 2>, <Codes.ERROR: 3>]
```

Enumerations can be created directly with sensible defaults,

```
>>> Codes = Enum('Lookup','START,STOP,ERROR')
>>> Codes.START
<Lookup.START: 1>
>>> Codes.ERROR
<Lookup.ERROR: 3>
```

You can look up the names corresponding to a value,

```
>>> Codes(1)
<Lookup.START: 1>
>>> Codes['START']
<Lookup.START: 1>
```

You can use the `unique` decorator to ensure that there are no duplicated values,

```
>>> from enum import unique
```

This will raise a `ValueError`

```
>>> @unique
... class UniqueCodes(Enum):
...     START = 1
...     END = 1 # same value as START
...
Traceback (most recent call last):
  File "<stdin>", line 2, in <module>
  File "/mnt/e/miniconda3/lib/python3.8/enum.py", line 860, in
  ↪  unique
    raise ValueError('duplicate values found in %r: %s' %
ValueError: duplicate values found in <enum 'UniqueCodes'>: END
  ↪   -> START
```

Type Annotations   Python 3 enables type annotations, which is a way to provide variable typing information for functions so that other tools like mypy can analyze large code bases to check for typing conflicts. This does *not* affect Python's dynamically typing. It means that in addition to creating unit tests, type annotations provide a supplemental way to improve code quality by uncovering defects *distinct* from those revealed by unit-testing.

```
# filename: type_hinted_001.py

def foo(fname:str) -> str:
    return fname+'.txt'

foo(10) # called with integer argument instead of string
```

Running this through mypy on the terminal command line by doing

```
% mypy type_hinted_001.py
```

will produce the following error:

```
type_hinted_001.py:6: error: Argument 1 to "foo" has incompatible
↪   type "int"; expected "str"
```

Functions that are not annotated (i.e., dynamically typed) are allowed in the same module as those that are annotated. The mypy might attempt to call out typing errors against these dynamically typed functions, but this behavior is considered unstable. You can supply both type annotations and default values, as in the following:

```
>>> def foo(fname:str = 'some_default_filename') -> str:
...     return fname+'.txt'
...
```

Types are *not* inferred from the types of the default values. The built-in typing module has definitions that can be used for type hinting,

```
>>> from typing import Iterable
>>> def foo(fname: Iterable[str]) -> str:
...     return "".join(fname)
...
```

The above declaration says that the input is an iterable (e.g., list) of strings and the function returns a single string as output. Since Python 3.6, you can also use type annotations for variables, as in the following,

```
>>> from typing import List
>>> x: str = 'foo'
>>> y: bool = True
>>> z: List[int] = [1]
```

Keep in mind that these additions are ignored in the interpreter and are processed by mypy separately. Type annotations also work with classes, as shown below,

```
>>> from typing import ClassVar
>>> class Foo:
...     count: ClassVar[int] = 4
...
```

Because type -hinting can become elaborate, especially with complex object patterns, type annotations can be segregated into pyi files. At this point, however, my feeling is that we have entered into a point of diminishing returns as the complexity burden is likely to overwhelm the benefits of maintaining such type checking. In the case of critical code that is maintained as separate functions utilizing primarily built-in Python types, then this additional burden may be warranted, but otherwise, and at least until the tooling required to aid the development and maintenance of type -annotations matures further, it is probably best to leave this aside.

Pathlib Python 3 has a new pathlib module that makes it easier to work with filesystems. Before pathlib, you had to use os.walk or another combination of directory searching tools to work with directories and paths. To get started, we import `Path` from the module,

```
>>> from pathlib import Path
```

The `Path` object itself provides useful information,

```
>>> Path.cwd() # gets current directory
PosixPath('/mnt/d/class_notes_pub')
>>> Path.home() # gets users home directory
PosixPath('/home/unpingco')
```

You can give the object a starting path,

```
>>> p = Path('./') # points to current directory
```

Then, you can search the directory path using its methods,

```
>>> p.rglob('*.log') # searches for log file extensions
<generator object Path.rglob at 0x7f939a5ecc80>
```

which returns a generator you can iterate through to obtain all `log` files in the named directory. Each returned element of the iteration is a `PosixPath` object with its own methods

```
>>> item = list(p.rglob('*.log'))[0]
>>> item
PosixPath('altair.log')
```

For example, the `stat` method provides the file metadata,

```
>>> item.stat()
os.stat_result(st_mode=33279, st_ino=3659174697287897, st_dev=15,
↪    st_nlink=1, st_uid=1000, st_gid=1000, st_size=52602,
↪    st_atime=1608339360, st_mtime=1608339360,
↪    st_ctime=1608339360)
```

The `pathlib` module can do much more such as creating files or processing elements of individual paths.

Asyncio   We have seen that generators/coroutines can exchange and act upon data so long as we manually drive this process. Indeed, one way to think about coroutines is as objects that require external management in order to do work. For generators, the business logic of the code usually manages the flow, but `async` simplifies this work and hides the tricky implementation details for doing this at large scale.

Consider the following function defined with the `async` keyword,

```
>>> async def sleepy(n=1):
...     print(f'n = {n} seconds')
...     return n
...
```

We can start this like we would a regular generator

```
>>> x = sleepy(3)
>>> type(x)
<class 'coroutine'>
```

Now, we get an event loop to drive this,

```
>>> import asyncio
>>> loop = asyncio.get_event_loop()
```

Now the `loop` can drive the coroutine,

```
>>> loop.run_until_complete(x)
n = 3 seconds
3
```

Let us put this in a synchronous *blocking* loop.

```
>>> from time import perf_counter
>>> tic = perf_counter()
>>> for i in range(5):
...     sleepy(0.1)
...
<coroutine object sleepy at 0x7f939a5f7ac0>
<coroutine object sleepy at 0x7f939a4027c0>
<coroutine object sleepy at 0x7f939a5f7ac0>
<coroutine object sleepy at 0x7f939a4027c0>
<coroutine object sleepy at 0x7f939a5f7ac0>
>>> print(f'elapsed time = {perf_counter()-tic}')
elapsed time = 0.0009255999993911246
```

Nothing happened! If we want to use the `sleepy` function inside other code, we need the `await` keyword,

```
>>> async def naptime(n=3):
...     for i in range(n):
...         print(await sleepy(i))
...
```

Roughly speaking, the `await` keywords means that the calling function should be suspended until the target of `await` has completed and control should be passed back to the event loop in the meantime. Next, we drive it as before with the event loop.

```
>>> loop = asyncio.get_event_loop()
>>> loop.run_until_complete(naptime(4))
n = 0 seconds
0
n = 1 seconds
1
n = 2 seconds
2
n = 3 seconds
3
```

Without the `await` keyword, the `naptime` function would just return the `sleepy` objects instead of the outputs from those objects. The function bodies *must* have asynchronous code in them or they will block. Python modules are developing that can play with this framework, but that is ongoing (e.g., `aiohttp` for asynchronous web access) at this time.

Let us consider another example that shows how the event loop is passed control for each of the asynchronous functions.

```
>>> async def task1():
...     print('entering task 1')
...     await asyncio.sleep(0)
...     print('entering task 1 again')
...     print('exiting task 1')
...
>>> async def task2():
...     print('passed into task2 from task1')
...     await asyncio.sleep(0)
...     print('leaving task 2')
...
>>> async def main():
...     await asyncio.gather(task1(),task2())
...
```

With all that set up, we have to kick this off with the event loop,

```
>>> loop = asyncio.get_event_loop()
>>> loop.run_until_complete(main())
entering task 1
passed into task2 from task1
entering task 1 again
exiting task 1
leaving task 2
```

The `await asyncio.sleep(0)` statement tells the event loop pass control to the next item (i.e., future or coroutine) because the current one is going to be busy waiting so the event loop might as well execute something else in the meantime. This means that tasks can finish out of order depending on how long they take when the event loop gets back to them. This next block demonstrates this.

```
>>> import random
>>> async def asynchronous_task(pid):
...     await asyncio.sleep(random.randint(0, 2)*0.01)
...     print('Task %s done' % pid)
...
>>> async def main():
...     await asyncio.gather(*[asynchronous_task(i) for i in
↪    range(5)])
...

>>> loop = asyncio.get_event_loop()
>>> loop.run_until_complete(main())
Task 0 done
Task 1 done
Task 2 done
Task 3 done
Task 4 done
```

This is all well and good from within the ecosystem of `asyncio`, but what to do for *existing* codes that are not so configured? We can use `concurrent.futures` to provide futures we can fold into the framework.

```
>>> from functools import wraps
>>> from time import sleep
>>> from concurrent.futures import ThreadPoolExecutor
>>> executor = ThreadPoolExecutor(5)
>>> def threadpool(f):
...        @wraps(f)
...        def wrap(*args, **kwargs):
...            return asyncio.wrap_future(executor.submit(f,
...                                                       *args,
...                                                       **kwargs))
...        return wrap
...
```

We can decorate a *blocking* version of `sleepy` as shown next and use `asyncio.wrap_future` to fold the thread into the `asyncio` framework,

```
>>> @threadpool
... def synchronous_task(pid):
...     sleep(random.randint(0, 1)) # blocking!
...     print('synchronous task %s done' % pid)
...
>>> async def main():
...     await asyncio.gather(synchronous_task(1),
...                          synchronous_task(2),
...                          synchronous_task(3),
...                          synchronous_task(4),
...                          synchronous_task(5))
...
```

With the following output. Note the non-ordering of the results.

```
>>> loop = asyncio.get_event_loop()
>>> loop.run_until_complete(main())
synchronous task 3 done
synchronous task 5 done
synchronous task 1 done
synchronous task 2 done
synchronous task 4 done
```

Debugging and Logging Python   The easiest way is to debug Python is on the command line,

```
% python -m pdb filename.py
```

and you will automatically enter the debugger. Otherwise, you can put `import pdb; pdb.set_trace()` source file where you want breakpoints. The advantage of using the semi-colon to put this all on one line is that it makes it easier to find and delete that one line later in your editor.

A cool trick that will provide a fully interactive Python shell anywhere in the code is to do the following:

```
import code; code.interact(local=locals());
```

Since Python 3.7, we have the `breakpoint` function, which is essentially the same as the `import pdb; pdb.set_trace()` line above but which allows for using the `PYTHONBREAKPOINT` environment variable to turn the breakpoint off or on, as in

```
% export PYTHONBREAKPOINT=0 python foo.py
```

In this case because the environment variable is set to zero, the `breakpoint` line in the source code of `foo.py` is ignored. Setting the environment variable to one will do the opposite. You can also choose your debugger using the environment variable. For example, if you prefer the `ipdb` IPython debugger, then you can do the following,

```
export PYTHONBREAKPOINT=ipdb.set_trace python foo.py
```

Using this environment variable, you can also have `breakpoint` run custom code when invoked. Given the following function,

```
# filename break_code.py
def do_this_at_breakpoint():
    print ('I am here in do_this_at_breakpoint')
```

Then, given we have `breakpoint` set in the `foo.py` file, we can do the following,

```
% export PYTHONBREAKPOINT=break_code.do_this_at_breakpoint python
↪   foo.py
```

and then the code will be run. Note that because this code is not invoking a debugger, the execution will not stop at the `breakpoint`. The `breakpoint` function can also take arguments,

```
breakpoint('a','b')
```

in your source code and then the invoking function will process those inputs, as in the following,

```
# filename break_code.py
def do_this_at_breakpoint(a,b):
    print ('I am here in do_this_at_breakpoint')
    print (f'argument a = {a}')
    print (f'argument b = {b}')
```

Then, the value of those variables at runtime will be printed out. Note that you can also explicitly invoke the debugger from within your custom breakpoint function by including the usual `import pdb; pdb.set_trace()` which will stop the code with the built-in debugger.

1.1.14 Using Python Assertions to Pre-debug Code

Asserts are a great way to provide sanity checks for your code. Using these is the quickest and easiest way to increase the *reliability* of your code! Note you can turn these off by running python on the command line with the `-O` option.

```
>>> import math
>>> def foo(x):
...       assert x>=0 # entry condition
...       return math.sqrt(x)
...
>>> foo(-1)
Traceback (most recent call last):
  File "<stdin>", line 1, in <module>
  File "<stdin>", line 2, in foo
AssertionError
```

Consider the following code,

```
>>> def foo(x):
...       return x*2
...
```

Any input x that understands the multiply operator will pass through. For example,

```
>>> foo('string')
'stringstring'
>>> foo([1,3,'a list'])
[1, 3, 'a list', 1, 3, 'a list']
```

which may not be what you want. To defend foo from this effect and enforce numeric input, you can do the following:

```
def foo(x):
    assert isinstance(x,(float,int,complex))
    return x*2
```

Or, even better, using abstract data types,

```
import numbers
def foo(x):
    assert isinstance(x,numbers.Number)
    return x*2
```

There is a philosophical argument about using assert in this situation because Python's duck-typing is supposed to reconcile this, but sometimes waiting for the traceback to complain is not feasible for your application. Assertions do much more than input type-checking. For example, suppose you have a problem where the sum of all the items in an intermediate list should add up to one. You can use assert in the code to verify this is true and then assert will automatically raise AssertionError if it is not.

1.1.15 Stack Tracing with sys.settrace

Although the pdb debugger is great for general purpose work, sometimes it 'is too painstaking to step through a long complicated program to find bugs. Python has a powerful tracing function that makes it possible to report every line of code that

Python executes and also filter those lines. Save the following code into a file and run it.

```python
# filename: tracer_demo.py
# demo tracing in python

def foo(x=10,y=10):
    return x*y

def goo(x,y=10):
    y= foo(x,y)
    return x*y

if __name__ =="__main__":
    import sys
    def tracer(frame,event,arg):
        if event=='line': # line about to be executed
            filename, lineno = frame.f_code.co_filename,
            ↪    frame.f_lineno
            print(filename,end='\t')                    # filename
            print(frame.f_code.co_name,end='\t')# function name
            print(lineno,end='\t')                      # line number in
            ↪    filename
            print(frame.f_locals,end='\t')     # local variables
            argnames =
            ↪    frame.f_code.co_varnames[:frame.f_code.co_argcount]
            print(' arguments:',end='\t')
            print(str.join(', ',['%s:%r' % (i,frame.f_locals[i]) for
            ↪    i in argnames]))
        return tracer # pass function along for next time

    sys.settrace(tracer)
    foo(10,30)
    foo(20,30)
    goo(33)
```

The key step is feeding the `tracer` function into `sys.settrace` which will run the `tracer` function on the stack frames and report out specified elements. You can also use the built-in tracer like

```
% python -m trace --ignore-module=sys --trace filename.py >
↪    output.txt
```

This will dump a *lot* of material so you probably want to use the other flags for filtering.

Programming Tip: Pysnooper
The tracing code above is helpful and a good place to start but the pynsooper module available on Pypi makes tracing *extremely* easy to implement and monitor and is highly recommended!

1.1.16 Debugging Using IPython

You can also use IPython from the command line as in

```
% ipython --pdb <filename>
```

You can also do this in the source code if you want to use the IPython debugger instead of the default.

```
from IPython.core.debugger import Pdb
pdb=Pdb() # create instance
for i in range(10):
    pdb.set_trace() # set breakpoint here
    print (i)
```

This provides the usual IPython dynamic introspection. This is compartmentalized into the ipdb package from PyPI. You can also invoke an embedded IPython shell by doing:

```
import IPython
for i in range(10):
    if i > 7:
        IPython.embed() # this will stop with an embedded IPython
        ↪ shell
    print (i)
```

This is handy when embedding Python in a GUI. Your mileage may vary otherwise, but it is a good trick in pinch!

1.1.17 Logging from Python

There is a powerful built-in logging package that beats putting print statements everywhere in your code, but it takes some setting up.

```
import logging, sys
logging.basicConfig(stream=sys.stdout, level=logging.INFO)
logging.debug("debug message") # no output
logging.info("info message") # output
logging.error("error")   # output
```

Note that the numerical values of the levels are defined as

```
>>> import logging
>>> logging.DEBUG
10
>>> logging.INFO
20
```

so when the logging level is set to INFO, only INFO and above messages are reported. You can also format the output using formatters.

```
import logging, sys
logging.basicConfig(stream=sys.stdout,
                    level=logging.INFO,
                    format="%(asctime)s - %(name)s -
                    ↪ %(levelname)s - %(message)s")

logging.debug("debug message") # no output
logging.info("info message")
logging.error("error")
```

So far, we have been using the *root* logger, but you can have many layers of organized logging based on the logger's name. Try running demo_log1.py in a console and see what happens

```
# top level program

import logging, sys
from demo_log2 import foo, goo

log = logging.getLogger('main') #name of logger
log.setLevel(logging.DEBUG)
handler = logging.StreamHandler(sys.stdout)
handler.setLevel(logging.DEBUG)

filehandler = logging.FileHandler('mylog.log')
formatter = logging.Formatter("%(asctime)s - %(name)s -
↪ %(funcName)s - %(levelname)s - %(message)s") # set format

handler.setFormatter(formatter) # setup format
filehandler.setFormatter(formatter) # setup format
log.addHandler(handler) # read to go
log.addHandler(filehandler) # read to go

def main(n=5):
    log.info('main called')
    [(foo(i),goo(i)) for i in range(n)]

if __name__ == '__main__':
    main()

# subordinate to demo_log1

import logging
log = logging.getLogger('main.demo_log2')

def foo(x):
    log.info('x=%r'%x)
    return 3*x

def goo(x):
    log = logging.getLogger('main.demo_log2.goo')
    log.info('x=%r'%x)
    log.debug('x=%r'%x)
    return 5*x**2

def hoo(x):
    log = logging.getLogger('main.demo_log2.hoo')
```

```
log.info('x=%r'%x)
return 5*x**2
```

Now, try changing the name of the `main` logger in `demo_log1.py` and see what happens. The logging in the `demo_log2.py` will *not* log unless it is subordinate to the `main` logger. You can see how embedding this in your code makes it easy to turn on various levels of code diagnosis. You can also have multiple handlers for different log levels.

Chapter 2
Object-Oriented Programming

Python is an object-oriented language. Object-oriented programming facilitates encapsulation of variables and functions and separates the various concerns of the program. This improves reliability because common functionality can be focused into the same code. In contrast with C++ or Java, you do not have to write custom classes in order to interact with Python's built-in features because object-oriented programming is not the only programming style that Python supports. In fact, I would argue that you want to wire together the built-in objects and classes that Python already provides as opposed to writing to own classes from scratch. In this section, we build up the background so you can write your own custom classes, if need be.

2.1 Properties/Attributes

The variables encapsulated by an object are called *properties* or *attributes*. Everything in Python is an object. For example,

```
>>> f = lambda x:2*x
>>> f.x = 30 # attach to x function object
>>> f.x
30
```

We have just hung an attribute on the function object. We can even reference it from within the function if we wanted.

```
>>> f = lambda x:2*x*f.x
>>> f.x = 30 # attach to x function object
>>> f.x
30
>>> f(3)
180
```

© The Editor(s) (if applicable) and The Author(s), under exclusive license to Springer Nature Switzerland AG 2021
J. Unpingco, *Python Programming for Data Analysis*,
https://doi.org/10.1007/978-3-030-68952-0_2

For safety reasons this arbitrary attaching of attributes to objects is blocked for certain built-in objects (c.f., __slots__), but you get the idea. You can create your own objects with the class keyword. The following is the simplest possible custom object,

```
>>> class Foo:
...      pass
...
```

We can instantiate our Foo object by calling it with parenthesis like a function in the following,

```
>>> f = Foo() # need parenthesis
>>> f.x = 30   # tack on attribute
>>> f.x
30
```

Note that we can attach our properties one-by-one as we did earlier with the built-in function object but we can use the __init__ method to do this for all instances of this class.

```
>>> class Foo:
...    def __init__(self): # note the double underscores
...         self.x = 30
...
```

The self keyword references the so-created instance. We can instantiate this object in the following,

```
>>> f = Foo()
>>> f.x
30
```

The __init__ constructor builds the attributes into all so-created objects,

```
>>> g = Foo()
>>> g.x
30
```

You can supply arguments to the __init__ function that are called upon instantiation,

```
>>> class Foo:
...      def __init__(self,x=30):
...          self.x = x
...
>>> f = Foo(99)
>>> f.x
99
```

Remember that the __init__ function is just another Python function and follows the same syntax. The surrounding double underscores indicate that the function has a special low-level status.

> **Programming Tip: Private vs. Public Attributes**
> Unlike many object-oriented languages, Python does *not* have to implement
> private versus public attributes as part of the language. Rather, these are
> managed by convention. For example, attributes that start with a single
> underscore character are (only by convention) considered *private*, although
> nothing in the language provides them special status.

2.2 Methods

Methods are functions that are attached to objects and have access to the internal
object attributes. They are defined within the body of the class definition,

```
>>> class Foo:
...        def __init__(self,x=30):
...            self.x = x
...        def foo(self,y=30):
...            return self.x*y
...
>>> f = Foo(9)
>>> f.foo(10)
90
```

Note that you can access the variables that were attached in `self.x` from within
the function body of `foo` with the `self` variables. A common practice in Python
coding is to pack all the non-changing variables in the attributes in the `__init__`
function and the set up the method so that the frequently changing variables are
then function variables, to be supplied by the user upon invocation. Also, `self`
can maintain state between method calls so that the object can maintain an internal
history and change the corresponding behavior of the object methods.

Importantly, methods always have at least one argument (i.e., `self`). For
example, see the following error,

```
>>> f.foo(10)     # this works fine
90
>>> f.foo(10,3) # this gives an error
Traceback (most recent call last):
  File "<stdin>", line 1, in <module>
TypeError: foo() takes from 1 to 2 positional arguments but 3
  were given
```

It looks like the method legitimately takes one argument for the first line, but why
does the error message say this method takes two arguments? The reason is that
Python sees `f.foo(10)` as `Foo.foo(f,10)` so the first argument is the instance
`f` which we referenced as `self` in the method definition. Thus, there are two
arguments from Python's perspective. This can be confusing the first time you see
it.

Programming Tip: Functions and Methods

Access to the attributes of the object is the distinction between methods and functions. For example, we can create a function and tack it onto an existing object, as in the following,

```
>>> f = Foo()
>>> f.func = lambda i:i*2
```

This is a legitimate function, and it can be called like a method f.func(10) but that function has *no* access to any of the internal attributes of f and must get all its arguments from the invocation.

Not surprisingly, methods can call other methods in the same object as long as they are referenced with the prefix self. Operations like the plus (+ operator) can be specified as methods.

```
>>> class Foo:
...     def __init__(self,x=10):
...         self.x = x
...     def __add__(self,y): # overloading "addition" operator
...         return self.x + y.x
...
>>> a=Foo(x=20)
>>> b=Foo()
>>> a+b
30
```

The operators module has a list of implemented operators.

Programming Tip: Calling Objects as Functions

Functions are objects in Python and you can make your class callable like a function by adding a __call__ method to your class. For example,

```
>>> class Foo:
...     def __call__(self,x):
...         return x*10
...
```

Now we can do something like,

```
>>> f = Foo()
>>> f(10)
100
```

The advantage of this technique is that now you can supply additional variables in the __init__ function and then just use the object like any other function.

2.3   Inheritance

Inheritance facilitates code-reuse. Consider the following,

```
>>> class Foo:
...     def __init__(self,x=10):
...         self.x = x
...     def compute_this(self,y=20):
...         return self.x*y
...
```

Now, let us suppose `Foo` works great except that we want to change the way
`compute_this` works for a new class. We do not have to rewrite the class, we can
just inherit from it and change (i.e., overwrite) the parts we do not like.

```
>>> class Goo(Foo): # inherit from Foo
...     def compute_this(self,y=20):
...         return self.x*y*1000
...
```

Now, this gives us everything in `Foo` except for the updated `compute_this`
function.

```
>>> g = Goo()
>>> g.compute_this(20)
200000
```

The idea is to reuse your own codes (or, better yet, someone else's) with inheritance.
Python also supports multiple inheritance and delegation (via the `super` keyword).

As an example, consider inheritance from the built-in `list` object where we
want to implement a special `__repr__` function.

```
>>> class MyList(list): # inherit from built-in list object
...     def __repr__(self):
...         list_string = list.__repr__(self)
...         return list_string.replace(' ','')
...
>>> MyList([1,3]) # no spaces in output
[1,3]
>>> list([1,3]) # spaces in output
[1, 3]
```

> **Programming Tip: The Advantages of a Good __repr__**
> The `repr` built-in function triggers the `__repr__` method, which is how
> the object is represented as a string. Strictly speaking, `repr` is supposed to
> return a string, when evaluated with the built-in `eval()` function, returns
> the an instance given object. Practically speaking, `repr` returns a string
> representation of the object and is an excellent opportunity to add useful
> psychologically advantageous notation to your objects, which makes your

(continued)

objects easier to reason about in the interactive interpreter or debugger. Here's an example of an object at represents an interval on the real line, which may be open or closed.

```
>>> class I:
...     def __init__(self,left,right,isopen=True):
...         self.left, self.right = left, right # edges of interval
...         self.isopen = isopen
...     def __repr__(self):
...         if self.isopen:
...             return '(%d,%d)'%(self.left,self.right)
...         else:
...             return '[%d,%d]'%(self.left,self.right)
...
>>> a = I(1,3) # open-interval representation?
>>> a
(1,3)
>>> b = I(11,13,False) # closed interval representation?
>>> b
[11,13]
```

Now it is visually obvious whether or not the given interval is open or closed by the enclosing parenthesis or square brackets. Providing this kind of psychological hint to yourself will make it much easier to reason about these objects.

Once you can write your own classes, you can reproduce the behavior of other Python objects, like iterables, for instance,

```
>>> class Foo:
...     def __init__(self,size=10):
...         self.size = size
...     def __iter__(self): # produces iterable
...         self.counter = list(range(self.size))
...         return self # return object that has next() method
...     def __next__(self): # does iteration
...         if self.counter:
...             return self.counter.pop()
...         else:
...             raise StopIteration
...
>>> f = Foo()
>>> list(f)
[9, 8, 7, 6, 5, 4, 3, 2, 1, 0]
>>> for i in Foo(3): # iterate over
...     print(i)
...
2
1
0
```

2.4 Class Variables

So far, we have been discussing properties and methods of objects as instances. You can specify variables tied to a class instead of an instance using *class variables*.

```
>>> class Foo:
...     class_variable = 10 # variables defined here are tied to
↪   the class not the particular instance
...
>>> f = Foo()
>>> g = Foo()
>>> f.class_variable
10
>>> g.class_variable
10
>>> f.class_variable = 20
>>> f.class_variable # change here
20
>>> g.class_variable # no change here
10
>>> Foo.class_variable  # no change here
10
>>> Foo.class_variable  = 100 # change this
>>> h = Foo()
>>> f.class_variable # no change here
20
>>> g.class_variable # change here even if pre-existing!
100
>>> h.class_variable # change here (naturally)
100
```

This also works with functions, not just variables, but only with the @classmethod decorator. Note that the existence of class variables does not make them known to the rest of the class definition automatically. For example,

```
>>> class Foo:
...     x = 10
...     def __init__(self):
...         self.fx = x**2 # x variable is not known
...
```

will result in the following error: NameError: global name 'x' is not defined. This can be fixed by providing the full class reference to x as in the following,

```
>>> class Foo:
...     x = 10
...     def __init__(self):
...         self.fx = Foo.x**2 # full reference to x
...
```

Although, it is probably best to avoid hard-coding the class name into the code, which makes downstream inheritance brittle.

2.5 Class Functions

Functions can be attached to classes also by using the `classmethod` decorator.

```
>>> class Foo:
...   @classmethod
...   def class_function(cls,x=10):
...       return x*10
...
>>> f = Foo()
>>> f.class_function(20)
200
>>> Foo.class_function(20) # don't need the instance
200
>>> class Foo:
...       class_variable = 10
...       @classmethod
...       def class_function(cls,x=10):
...           return x*cls.class_variable #using class_variable
...
>>> Foo.class_function(20) # don't need the instance, using the
↪    class_variable
200
```

This can be useful if you want to transmit a parameter to all class instances after construction. For example,

```
>>> class Foo:
...       x=10
...       @classmethod
...       def foo(cls):
...             return cls.x**2
...
>>> f = Foo()
>>> f.foo()
100
>>> g = Foo()
>>> g.foo()
100
>>> Foo.x = 100 # change class variable
>>> f.foo() # now the instances pickup the change
10000
>>> g.foo() # now the instances pickup the change
10000
```

This can be tricky to keep track of because the class itself holds the class variable. For example:

```
>>> class Foo:
...       class_list = []
...       @classmethod
...       def append_one(cls):
...           cls.class_list.append(1)
...

>>> f = Foo()
>>> f.class_list
```

```
[]
>>> f.append_one()
>>> f.append_one()
>>> f.append_one()
>>> g = Foo()
>>> g.class_list
[1, 1, 1]
```

Notice how the new object g got the changes in the class variable that we made by the f instance. Now, if we do the following:

```
>>> del f,g
>>> Foo.class_list
[1, 1, 1]
```

Note that the class variable is attached to the class definition so deleting the class instances did *not* affect it. Make sure this is the behavior you expect when you set up class variables this way!

Class variables and methods are *not* lazy evaluated. This will become apparent when we get to dataclasses. For example,

```
>>> class Foo:
...     print('I am here!')
...
I am here!
```

Notice that we did *not* have to instantiate an instance of this class in order to execute the print statement. This has important subtleties for object-oriented design. Sometimes platform-specific parameters are inserted as class variables so they will be set up by the time any instance of the class is instantiated. For example,

```
>>> class Foo:
...     _setup_const = 10 # get platform-specific info
...     def some_function(self,x=_setup_const):
...         return 2*x
...
>>> f = Foo()
>>> f.some_function()
20
```

2.6   Static Methods

As opposed to class methods, a staticmethod is attached to the class definition but does not need access to the internal variables. Consider the following:

```
>>> class Foo:
...     @staticmethod
...     def mystatic(x,y):
...         return x*y
...
```

The staticmethod does not have access to the internal self or cls that a regular method instance or classmethod has, respectively. It is just a way to attach a function to a class. Thus,

```
>>> f = Foo()
>>> f.mystatic(1,3)
3
```

Sometimes you will find these in mix-in objects, which are objects that are designed to not touch any of the internal self or cls variables.

2.7　Hashing Hides Parent Variables from Children

By convention, methods and attributes that start with a single underscore character are considered *private* but those that start with a double underscore are internally *hashed* with the class name,

```
>>> class Foo:
...      def __init__(self):
...          self.__x=10
...      def count(self):
...          return self.__x*30
...
```

Note that the count function utilizes the double underscored variable self.__x.

```
>>> class Goo(Foo): # child with own .__x attribute
...      def __init__(self,x):
...          self.__x=x
...
>>> g=Goo(11)
>>> g.count() # won't work
Traceback (most recent call last):
  File "<stdin>", line 1, in <module>
  File "<stdin>", line 5, in count
AttributeError: 'Goo' object has no attribute '_Foo__x'
```

This means that the __x variable in the Foo declaration is attached to the Foo class. This is to prevent a potential subclass from using the Foo.count() function with a subclass's variable (say, self._x, without the double underscore).

2.8　Delegating Functions

In the following, try to think through where the abs function is originating from in the chain of inheritance below.

```
>>> class Foo:
...      x = 10
```

```
...      def abs(self):
...           return abs(self.x)
...
>>> class Goo(Foo):
...      def abs(self):
...           return abs(self.x)*2
...
>>> class Moo(Goo):
...      pass
...
>>> m = Moo()
>>> m.abs()
20
```

When Python sees m.abs() it first checks to see if the Moo class implements the abs() function. Then, because it does not, it reads from left to right in the inheritance and finds Goo. Because Goo *does* implement the abs() function it uses this one but it requires self.x which Goo gets from its parent Foo in order to finish the calculation. The abs() function in Goo relies upon the built-in Python abs() function.

2.9 Using super for Delegation

Python's super method is a way to run functions along the class' Method Resolution Order (MRO). A better name for super would be next-method-in-MRO. In the following, both classes A and B inherit from Base,

```
>>> class Base:
...      def abs(self):
...           print('in Base')
...           return 1
...
>>> class A(Base):
...      def abs(self):
...           print('in A.abs')
...           oldval=super(A,self).abs()
...           return abs(self.x)+oldval
...
>>> class B(Base):
...      def abs(self):
...           print('in B.abs')
...           oldval=super(B,self).abs()
...           return oldval*2
...
```

With all of that set up, let us create a new class that inherits from both A and B with the inheritance tree in Fig. 2.1,

```
>>> class C(A,B):
...      x=10
...      def abs(self):
```

```
...             return super(C,self).abs()
...
>>> c=C() # create an instance of class C
>>> c.abs()
in A.abs
in B.abs
in Base
12
```

What happened? As shown in Fig. 2.1, the method resolution looks for an `abs` function in the class C, finds it, and then goes to the next method in the order of the inheritance (class A). It finds an `abs` function there and executes it and then moves on to the next class in the MRO (class B) and then also finds an `abs` function there and subsequently executes it. Thus, `super` basically daisy-chains the functions together.

Now let us see what happens when we change the method resolution order by starting at class A as in the following:

```
>>> class C(A,B):
...     x=10
...     def abs(self):
...             return super(A,self).abs()
...
>>> c=C() # create an instance of class C
>>> c.abs()
in B.abs
in Base
2
```

Fig. 2.1 Method resolution order

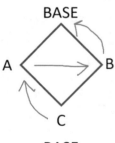

Fig. 2.2 Method resolution order

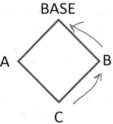

Notice that it picks up the method resolution after class A (see Fig. 2.2). We can change the order of inheritance and see how this affects the resolution order for super.

```
>>> class C(B,A): # change MRO
...     x=10
...     def abs(self):
...         return super(B,self).abs()
...
>>> c=C()
>>> c.abs()
in A.abs
in Base
11

>>> class C(B,A): # same MRO, different super
...     x=10
...     def abs(self):
...         return super(C,self).abs()
...
>>> c=C()
>>> c.abs()
in B.abs
in A.abs
in Base
22
```

To sum up, super allows you to mix and match objects to come up with different usages based upon how the method resolution is resolved for specific methods. This adds another degree of freedom but also another layer of complexity to your code.

2.10 Metaprogramming: Monkey Patching

Monkey Patching means adding methods to classes or instances outside of the formal class definition. This makes it very hard to debug, because you do not know exactly where functions have been hijacked. Although highly inadvisable, you can monkey patch class methods also:

```
>>> import types
>>> class Foo:
...     class_variable = 10
...
>>> def my_class_function(cls,x=10):
...     return x*10
...
>>> Foo.my_class_function = types.MethodType(my_class_function,
↪    Foo)
>>> Foo.my_class_function(10)
100
>>> f = Foo()
>>> f.my_class_function(100)
1000
```

This can be useful for last-minute debugging

```
>>> import types
>>> class Foo:
...     @classmethod
...     def class_function(cls,x=10):
...         return x*10
...
>>> def reported_class_function():
...     # hide original function in a closure
...     orig_function = Foo.class_function
...     # define replacement function
...     def new_class_function(cls,x=10):
...         print('x=%r' % x)
...         return orig_function(x)# return using original function
...     return new_class_function # return a FUNCTION!
...
>>> Foo.class_function(10) # original method
100
>>> Foo.class_function = types.MethodType(reported_class_
function(), Foo)
>>> Foo.class_function(10) # new verbose method
x=10
100
```

Programming Tip: Keep It Simple
It is easy to get fancy with Python classes, but it is best to use classes and object-oriented programming where they conceptually simplify and unify codes, avoid code redundancy, and maintain organizational simplicity. To see a perfect application of Python object-oriented design, study the Networkx [2] graph module.

2.11 Abstract Base Classes

The collections module contains Abstract Base Classes which serve two primary functions. First, they provide a way to check if a given custom object has a desired interface using isinstance or issubclass. Second, they provide a minimum set of requirements for new objects in order to satisfy specific software patterns. Let us consider a function such as g = lambda x:x**2. Suppose we want to test if g is a *callable*. One way to check this is to use the callable function as in

```
>>> g = lambda x:x**2
>>> callable(g)
True
```

Alternatively, using Abstract Base Classes, we can do the same test as in the following:

```
>>> from collections.abc import Callable
>>> isinstance(g,Callable)
True
```

The Abstract Base Classes extend this capability to the given interfaces as described in the main Python documentation. For example, to check whether or not a custom object is iterable, we can do the following:

```
>>> from collections.abc import Iterable
>>> isinstance(g,Iterable)
False
```

Beyond this kind of interface checking, Abstract Base Classes also allow object designers to specify a minimum set of methods and to get the remainder of the methods that characterize a particular Abstract Base Class. For example, if we want to write a dictionary-like object, we can inherit from the MutableMapping Abstract Base Class and then write the __getitem__, __setitem__, __delitem__, __iter__, __len__ methods. Then, we get the other MutableMapping methods like clear(), update(), etc. for free as part of the inheritance.

ABCMeta Programming The machinery that implements the above metaclasses is also available directly via the abc module. Abstract Base Classes can also enforce subclass methods by using the abc.abstractmethod decorator. For example,

```
>>> import abc
>>> class Dog(metaclass=abc.ABCMeta):
...     @abc.abstractmethod
...     def bark(self):
...         pass
...
```

This means that all the subclasses of Dog have to implement a bark method or TypeError will be thrown. The decorator marks the method as *abstract*.

```
>>> class Pug(Dog):
...     pass
...
>>> p = Pug() # throws a TypeError
Traceback (most recent call last):
  File "<stdin>", line 1, in <module>
TypeError: Can't instantiate abstract class Pug with abstract
 ↪   methods bark
```

So you must implement the bark method in the subclass,

```
>>> class Pug(Dog):
...     def bark(self):
...         print('Yap!')
...
>>> p = Pug()
```

Then,

```
>>> p.bark()
Yap!
```

Besides subclassing from the base class, you can also use the `register` method
to take another class and make it a subclass, assuming it implements the desired
abstract methods, as in the following:

```
>>> class Bulldog:
...     def bark(self):
...         print('Bulldog!')
...
>>> Dog.register(Bulldog)
<class '__console__.Bulldog'>
```

Then, even though `Bulldog` is not written as a subclass of `Dog`, it will still act that
way:

```
>>> issubclass(Bulldog, Dog)
True
>>> isinstance(Bulldog(), Dog)
True
```

Note that without the `bark` method, we would *not* get an error if we tried to
instantiate the `Bulldog` class.

Even though the concrete implementation of the abstract method is the responsi-
bility of the subclass writer, you can still user `super` to run the main definition in
the parent class. For example,

```
>>> class Dog(metaclass=abc.ABCMeta):
...     @abc.abstractmethod
...     def bark(self):
...         print('Dog bark!')
...
>>> class Pug(Dog):
...     def bark(self):
...         print('Yap!')
...         super(Pug,self).bark()
...
```

Then,

```
>>> p= Pug()
>>> p.bark()
Yap!
Dog bark!
```

2.12 Descriptors

Descriptors expose the internal abstractions inside of the Python object-creation
technology for more general usage. The easiest way to get started with descriptors
is to hide input validation in an object from the user. For example:

```
>>> class Foo:
...     def __init__(self,x):
...         self.x = x
...
```

This object has a simple attribute called x. There is nothing stopping the user from doing the following:

```
>>> f = Foo(10)
>>> f.x = 999
>>> f.x = 'some string'
>>> f.x = [1,3,'some list']
```

In other words, nothing is stopping the attribute x from being assigned to all of these different types. This might not be what you want to do. If you want to ensure that this attribute is only ever assigned an integer, for example, you need a way to enforce that. That is where descriptors come in. Here is how:

```
>>> class Foo:
...     def __init__(self,x):
...         assert isinstance(x,int) # enforce here
...         self._x = x
...     @property
...     def x(self):
...         return self._x
...     @x.setter
...     def x(self,value):
...         assert isinstance(value,int) # enforce here
...         self._x = value
...
```

The interesting part happens with the `property` decorator. This is where we explicitly change how Python deals with the x attribute. From now on, when we try to set the value of this attribute instead of going directly to the `Foo.__dict__`, which is the object dictionary that holds all of the attributes of the `Foo` object, the `x.setter` function will be called instead and the assignment will be handled there. Analogously, this also works for retrieving the value of the attribute, as shown below with the `x.getter` decorator:

```
>>> class Foo:
...     def __init__(self,x):
...         assert isinstance(x,int) # enforce here
...         self._x = x
...     @property
...     def x(self):
...         return self._x
...     @x.setter
...     def x(self,value):
...         assert isinstance(value,int) # enforce here
...         self._x = value
...     @x.getter
...     def x(self):
...         print('using getter!')
...         return self._x
...
```

The advantage of this `@getter` technique is that now we can return a computed value every time the attribute is accessed. For example, we could do something like the following:

```
@x.getter
def x(self):
   return self._x * 30
```

So now, unbeknownst to the user, the attribute is computed dynamically. Now, here comes the punchline: the entire descriptor machinery can be abstracted away from the class definition. This is really useful when the same set of descriptors have to be reused multiple times within the same class definition and rewriting them one-by-one for every attribute would be error-prone and tedious. For example, suppose we have the following class for `FloatDescriptor` and then another class that describes a `Car`.

```
>>> class FloatDescriptor:
...        def __init__(self):
...            self.data = dict()
...        def __get__(self, instance, owner):
...            return self.data[instance]
...        def __set__(self, instance, value):
...            assert isinstance(value,float)
...            self.data[instance] = value
...
>>> class Car:
...     speed = FloatDescriptor()
...     weight =  FloatDescriptor()
...     def __init__(self,speed,weight):
...        self.speed = speed
...        self.weight = weight
...
```

Note that FloatDescriptor appears as class variables in the class definition of Car. This will have important ramifications later, but for now let us see how this works.

```
>>> f = Car(1,2) # raises AssertionError
Traceback (most recent call last):
  File "<stdin>", line 1, in <module>
  File "<stdin>", line 5, in __init__
  File "<stdin>", line 7, in __set__
AssertionError
```

This is because the descriptor demands float parameters.

```
>>> f = Car(1.0,2.3) # no AssertionError because FloatDescriptor
↪  is satisfied
>>> f.speed = 10        # raises AssertionError
Traceback (most recent call last):
  File "<stdin>", line 1, in <module>
  File "<stdin>", line 7, in __set__
AssertionError
>>> f.speed = 10.0    # no AssertionError
```

Now that we have abstracted away the management and validation of the attributes of the Car class using FloatDescriptor and we can then reuse FloatDescriptor in other classes. However, there is a big caveat here because we had to use FloatDescriptor at the class level in order to get the descriptors hooked in correctly. This means that we have to ensure that the assignment of the instance attributes is placed on the correct instance. This is why `self.data` is a dictionary in the `FloatDescriptor` constructor. We are using the instance itself as the key to this dictionary in order to ensure that the attributes get placed on the correct instance, as in the following: `self.data[instance] = value`. This can fail for classes that are non-hashable and that therefore cannot be used as dictionary keys. The reason the `__get__` has an `owner` argument is that these issues can be resolved using metaclasses, but that is far outside of our scope.

To sum up, descriptors are the low-level mechanism that Python classes use internally to manage methods and attributes. They also provide a way to abstract the management of class attributes into separate descriptor classes that can be shared between classes. Descriptors can get tricky for non-hashable classes and there are other issues with extending this pattern beyond what we have discussed here. The book *Python Essential Reference* [1] is an excellent reference for advanced Python.

2.13   Named Tuples and Data Classes

Named tuples allow for easier and more readable access to tuples. For example,

```
>>> from collections import namedtuple
>>> Produce = namedtuple('Produce','color shape weight')
```

We have just created a new class called `Produce` that has the attributes `color`, `shape`, and `weight`. Note that you cannot have Python keywords or duplicated names in the attributes specification. To use this new class, we just instantiate it like any other class,

```
>>> mango = Produce(color='g',shape='oval',weight=1)
>>> print (mango)
Produce(color='g', shape='oval', weight=1)
```

Note that we can get at the elements of the tuple by using the usual indexing,

```
>>> mango[0]
'g'
>>> mango[1]
'oval'
>>> mango[2]
1
```

We can get the same by using the named attributes,

```
>>> mango.color
'g'
>>> mango.shape
'oval'
>>> mango.weight
1
```

Tuple unpacking works as with regular tuples,

```
>>> i,j,k = mango
>>> i
'g'
>>> j
'oval'
>>> k
1
```

You can get the names of the attributes also,

```
>>> mango._fields
('color', 'shape', 'weight')
```

We can create new namedtuple objects by replacing values of the existing attributes with the _replace method, as in the following,

```
>>> mango._replace(color='r')
Produce(color='r', shape='oval', weight=1)
```

Under the hood, namedtuple automatically generates code to implement the corresponding class (Produce in this case). This idea of auto-generating code to implement specific classes is extended with dataclasses in Python 3.7$^+$.

Data Classes In Python 3.7$^+$, dataclasses extend the code generation idea beyond namedtuple to more generic data-like objects.

```
>>> from dataclasses import dataclass
>>> @dataclass
... class Produce:
...     color: str
...     shape: str
...     weight: float
...
>>> p = Produce('apple','round',2.3)
>>> p
Produce(color='apple', shape='round', weight=2.3)
```

Do not be tricked into thinking that the given types are enforced,

```
>>> p = Produce(1,2,3)
>>> p
Produce(color=1, shape=2, weight=3)
```

We get many additional methods for free using the dataclass decorator,

```
>>> dir(Produce)
['__annotations__', '__class__', '__dataclass_fields__',
'__dataclass_params__', '__delattr__', '__dict__', '__dir__',
'__doc__', '__eq__', '__format__', '__ge__', '__getattribute__',
'__gt__', '__hash__', '__init__', '__init_subclass__', '__le__',
'__lt__', '__module__', '__ne__', '__new__', '__reduce__',
'__reduce_ex__', '__repr__', '__setattr__', '__sizeof__',
'__str__', '__subclasshook__', '__weakref__']
```

The __hash__() and __eq__() are particularly useful for allowing these objects to be used as keys in a dictionary but you have to use the frozen=True keyword argument as shown,

```
>>> @dataclass(frozen=True)
... class Produce:
...     color: str
...     shape: str
...     weight: float
...
>>> p = Produce('apple','round',2.3)
>>> d = {p: 10} # instance as key
>>> d
{Produce(color='apple', shape='round', weight=2.3): 10}
```

You can also use `order=True` if you want the class to order based on the tuple of inputs. Default values can be assigned as in the following,

```
>>> @dataclass
... class Produce:
...     color: str = 'red'
...     shape: str = 'round'
...     weight: float = 1.0
...
```

Unlike `namedtuple`, you can have custom methods,

```
>>> @dataclass
... class Produce:
...     color  : str
...     shape  : str
...     weight : float
...     def price(self):
...         return  0 if self.color=='green' else self.weight*10
...
```

Unlike `namedtuple`, `dataclass` is not iterable. There are helper functions that can be used. The `field` function allows you to specify how certain declared attributes are created by default. The example below uses a `list` factory to avoid all the instances of the class sharing the same `list` as a class variable.

```
>>> from dataclasses import field
>>> @dataclass
... class Produce:
...     color  : str = 'green'
...     shape  : str = 'flat'
...     weight : float = 1.0
...     track  : list = field(default_factory=list)
...
```

Thus, two different instances of `Produce` have different mutable `track` lists. This avoids the problem of using a mutable object in the initializer. Other arguments for `dataclass` allow you to automatically define an order on your objects or make them immutable, as in the following.

```
>>> @dataclass(order=True,frozen=True)
... class Coor:
...     x: float = 0
...     y: float = 0
```

```
...
>>> c = Coor(1,2)
>>> d = Coor(2,3)
>>> c < d
True
>>> c.x = 10
Traceback (most recent call last):
  File "<stdin>", line 1, in <module>
  File "<string>", line 4, in __setattr__
dataclasses.FrozenInstanceError: cannot assign to field 'x'
```

The `asdict` function easily converts your dataclasses to regular Python dictionaries, which is useful for serialization. Note that this will only convert the *attributes* of the instance.

```
>>> from dataclasses import asdict
>>> asdict(c)
{'x': 1, 'y': 2}
```

If you have variables that are dependent on other initialized variables, but you do not want to auto-create them with every new instance, then you can use the `field` function, as in the following,

```
>>> @dataclass
... class Coor:
...     x : float = 0
...     y : float = field(init=False)
...
>>> c = Coor(1) # y is not specified on init
>>> c.y = 2*c.x # added later
>>> c
Coor(x=1, y=2)
```

That is a little cumbersome and is the reason for the `__post_init__` method. Remember that the `__init__` method is auto-generated by `dataclass`.

```
>>> @dataclass
... class Coor:
...     x : float = 0
...     y : float = field(init=False)
...     def __post_init__(self):
...         self.y = 2*self.x
...
>>> c = Coor(1) # y is not specified on init
>>> c
Coor(x=1, y=2)
```

To sum up, dataclasses are new and it remains to be seen how they will ultimately fit into common workflows. These dataclasses are inspired by the third-party `attrs` module, so read up on that module to understand if their use cases apply to your problems.

2.14   Generic Functions

Generic functions are those that change their implementations based upon the types of the inputs. For example, you could accomplish the same thing using the following conditional statement at the start of a function as shown in the following,

```
>>> def foo(x):
...     if isinstance(x,int):
...         return 2*x
...     elif isinstance(x,list):
...         return [i*2 for i in x]
...     else:
...         raise NotImplementedError
...
```

With the following usage,

```
>>> foo(1)
2
>>> foo([1,2])
[2, 4]
```

In this case, you can think of foo as a *generic* function. To put more reliability behind this pattern, since Python 3.3, we have functools.singledispatch. To start, we need to define the top level function that will template-out the individual implementations based upon the type of the *first* argument.

```
>>> from functools import singledispatch
>>> @singledispatch
... def foo(x):
...     print('I am done with type(x): %s'%(str(type(x))))
...
```

With the corresponding output,

```
>>> foo(1)
I am done with type(x): <class 'int'>
```

To get the dispatch to work we have to register the new implementations with foo using the type of the input as the argument to the decorator. We can name the function _ because we do not need a separate name for it.

```
>>> @foo.register(int)
... def _(x):
...     return 2*x
...
```

Now, let us try the output again and notice that the new int version of the function has been executed.

```
>>> foo(1)
2
```

We can tack on more type-based implementations by using register again with different type arguments,

```
>>> @foo.register(float)
... def _(x):
...     return 3*x
...
>>> @foo.register(list)
... def _(x):
...     return [3*i for i in x]
...
```

With the corresponding output,

```
>>> foo(1.3)
3.9000000000000004
>>> foo([1,2,3])
[3, 6, 9]
```

Existing functions can be attached using the functional form of the decorator,

```
>>> def existing_function(x):
...     print('I am the existing_function with %s'%(str(type(x))))
...
>>> foo.register(dict,existing_function)
<function existing_function at 0x7f9398354a60>
```

With the corresponding output,

```
>>> foo({1:0,2:3})
I am the existing_function with <class 'dict'>
```

You can see the implemented dispatches using `foo.registry.keys()`, as in the following,

```
>>> foo.registry.keys()
dict_keys([<class 'object'>, <class 'int'>, <class 'float'>,
 <class 'list'>, <class 'dict'>])
```

You can pick out the individual functions by accessing the dispatch, as in the following,

```
>>> foo.dispatch(int)
<function _ at 0x7f939a66b5e0>
```

These `register` decorators can also be stacked and used with Abstract Base Classes.

> **Programming Tip: Using Slots to Cut Down Memory**
> The following class definition allows arbitrary addition of attributes.
>
> ```
> >>> class Foo:
> ...     def __init__(self,x):
> ...         self.x=x
> ...
> ```

(continued)

```
>>> f = Foo(10)
>>> f.y = 20
>>> f.z = ['some stuff', 10,10]
>>> f.__dict__
{'x': 10, 'y': 20, 'z': ['some stuff', 10, 10]}
```

This is because there is a dictionary inside Foo, which creates memory overhead, especially for many such objects (see prior discussion of garbage collection). This overhead can be removed by adding __slots__ as in the following,

```
>>> class Foo:
...     __slots__ = ['x']
...     def __init__(self,x):
...         self.x=x
...
>>> f = Foo(10)
>>> f.y = 20
Traceback (most recent call last):
  File "<stdin>", line 1, in <module>
AttributeError: 'Foo' object has no attribute 'y'
```

This raises AttributeError because __slots__ prevents Python from creating an internal dictionary for each instance of Foo.

2.15 Design Patterns

Design Patterns are not as popular in Python as opposed to Java or C^{++} because Python has such a wide-ranging and useful standard library. Design Patterns represent canonical solutions to common problems. The terminology derives from architecture. For example, suppose you have a house and your problem is how to enter the house while carrying a bag of groceries. The solution to this problem is the *door pattern*, but this does not specify the shape or size of the door, its color, or whether or not it has a lock on it, etc. These are known as *implementation details*. The main idea is that there are canonical solutions for common problems.

2.15.1 Template

The Template Method is a behavioral design pattern that defines the skeleton of an algorithm in the base class but lets subclasses override specific steps of the algorithm without changing its structure. The motivation is to break an algorithm into a series of steps where each step is has an abstract implementation. Specifically, this means

that the implementation details of the algorithm are left to the subclass while the
base class orchestrates the individual steps of the algorithm.

```
>>> from abc import ABC, abstractmethod
>>> # ensures it must be subclassed not directly instantiated
>>> class Algorithm(ABC):
...      # base class method
...      def compute(self):
...          self.step1()
...          self.step2()
...      # subclasses must implement these abstractmethods
...      @abstractmethod
...      def step1(self):
...          'step 1 implementation details in subclass'
...          pass
...      @abstractmethod
...      def step2(self):
...          'step 2 implementation details in subclass'
...          pass
...
```

Python throws a `TypeError` if you try to instantiate this object directly. To use the
class, we have to subclass it as in the following,

```
>>> class ConcreteAlgorithm(Algorithm):
...      def step1(self):
...          print('in step 1')
...      def step2(self):
...          print('in step 2')
...
>>> c = ConcreteAlgorithm()
>>> c.compute() # compute is defined in base class
in step 1
in step 2
```

The advantage of the template pattern is that it makes it clear that the base class
coordinates the details that are implemented by the subclasses. This separates the
concerns clearly and makes it flexible to deploy the same algorithm in different
situations.

2.15.2 Singleton

Singleton is a creational design pattern to guarantee that a class has only one
instance. For example, there can be many printers but only *one* printer spooler. The
following code hides the singular instance in the class variable and uses the __new__
method to customize object creation before __init__ is called.

```
>>> class Singleton:
...      # class variable contains singular _instance
...      # __new__ method returns object of specified class and is
...      #    called before __init__
```

```
...        def __new__(cls, *args, **kwds):
...            if not hasattr(cls, '_instance'):
...                cls._instance = super().__new__(cls, *args, **kwds)
...            return cls._instance
...
>>> s = Singleton()
>>> t = Singleton()
>>> t is s # there can be only one!
True
```

Note that there are many ways to implement this pattern in Python, but this is one of the simplest.

2.15.3   Observer

Observer is a behavioral design pattern that defines communication between objects so that when one object (i.e., publisher) changes state, all its subscribers are notified and updated automatically. The traitlets module implements this design pattern.

```
>>> from traitlets import HasTraits, Unicode, Int
>>> class Item(HasTraits):
...        count = Int()    # publisher for integer
...        name = Unicode() # publisher for unicode string
...
>>> def func(change):
...        print('old value of count = ',change.old)
...        print('new value of count = ',change.new)
...
>>> a = Item()
>>> # func subscribes to changes in `count`
>>> a.observe(func, names=['count'])
>>> a.count = 1
old value of count =   0
new value of count =   1
>>> a.name = 'abc'  # prints nothing but not watching name
```

With all that set up, we can have multiple subscribers to published attributes

```
>>> def another_func(change):
...        print('another_func is subscribed')
...        print('old value of count = ',change.old)
...        print('new value of count = ',change.new)
...
>>> a.observe(another_func, names=['count'])
>>> a.count = 2
old value of count =   1
new value of count =   2
another_func is subscribed
old value of count =   1
new value of count =   2
```

Additionally, the `traitlets` module does type-checking of the object attributes that will raise an exception if the wrong type is set to the attribute, implementing the `descriptor` pattern. The `traitlets` module is fundamental to the interactive web-based features of the Jupyter `ipywidgets` ecosystem.

2.15.4 Adapter

The Adapter patterns facilitate reusing existing code by impersonating the relevant interfaces using classes. This permits classes to interoperate that otherwise could not due to incompatibilities in their interfaces. Consider the following class that takes a list,

```
>>> class EvenFilter:
...     def __init__(self,seq):
...         self._seq = seq
...     def report(self):
...         return [i for i in self._seq if i%2==0]
...
```

This returns only the even terms as shown below,

```
>>> EvenFilter([1,3,4,5,8]).report()
[4, 8]
```

But now we want to use the same class where the input `seq` is now a generator instead of a list. We can do this with the following `GeneratorAdapter` class,

```
>>> class GeneratorAdapter:
...     def __init__(self,gen):
...         self._seq = list(gen)
...     def __iter__(self):
...         return iter(self._seq)
...
```

Now we can go back and use this with the `EvenFilter` as follows:

```
>>> g = (i for i in range(10)) # create generator with
↪   comprehension
>>> EvenFilter(GeneratorAdapter(g)).report()
[0, 2, 4, 6, 8]
```

The main idea for the adapter pattern is to isolate the relevant interfaces and impersonate them with the adapter class.

References

1. D.M. Beazley, *Python Essential Reference* (Addison-Wesley, Boston, 2009)
2. A.A. Hagberg, D.A. Schult, P.J. Swart, Exploring network structure, dynamics, and function using NetworkX, in *Proceedings of the 7th Python in Science Conference (SciPy2008)*, Pasadena, CA, August 2008, pp. 11–15

Chapter 3
Using Modules

Modules allow for code-reuse and portability. It is generally better to use widely used and heavily tested code than to write your own from scratch. The `import` statement is how modules are loaded into the current namespace. Behind the scenes, importing is a complex process. Consider the following import statement:

```
import some_module
```

To import this module, Python will search for a valid Python module in the order of the entries in the `sys.path` directory list. The items in the `PYTHONPATH` environment variable are added to this search path. How Python was compiled affects the import process. Generally speaking, Python searches the filesystem for modules, but certain modules may be compiled into Python directly, meaning that it knows where to load them without searching the filesystem. This can have massive performance consequences when starting thousands of Python process on a shared filesystem because the filesystem may cause significant startup delays as it is pounded during searching.

3.1 Standard Library

Python is a *batteries included* language, meaning that lots of excellent modules are already included in the base language. Due to its legacy as a web programming language, most of the standard libraries deal with network protocols and other topics important to web development. The standard library modules are documented on the main Python site. Let us look at the built-in `math` module,

```
>>> import math        # Importing math module
>>> dir(math)          # Provides a list of module attributes
['__doc__', '__file__', '__loader__', '__name__', '__package__',
'__spec__', 'acos', 'acosh', 'asin', 'asinh', 'atan', 'atan2',
'atanh', 'ceil', 'comb', 'copysign', 'cos', 'cosh', 'degrees',
```

© The Editor(s) (if applicable) and The Author(s), under exclusive license
to Springer Nature Switzerland AG 2021
J. Unpingco, *Python Programming for Data Analysis*,
https://doi.org/10.1007/978-3-030-68952-0_3

```
'dist', 'e', 'erf', 'erfc', 'exp', 'expm1', 'fabs', 'factorial',
'floor', 'fmod', 'frexp', 'fsum', 'gamma', 'gcd', 'hypot', 'inf',
'isclose', 'isfinite', 'isinf', 'isnan', 'isqrt', 'ldexp',
'lgamma', 'log', 'log10', 'log1p', 'log2', 'modf', 'nan', 'perm',
'pi', 'pow', 'prod', 'radians', 'remainder', 'sin', 'sinh',
'sqrt', 'tan', 'tanh', 'tau', 'trunc']
>>> help(math.sqrt)
Help on built-in function sqrt in module math:

sqrt(x, /)
    Return the square root of x.

>>> radius = 14.2
>>> area = math.pi*(radius**2)
>>> area                        # Using a module variable
633.4707426698459
>>> a = 14.5; b = 12.7
>>> c = math.sqrt(a**2+b**2) # Using a module function
>>> c
19.275372888740698
```

Once a module is imported, it is included in the sys.modules dictionary. The
first step of import is to check this dictionary for the desired module and not
re-import it again. This means that doing import somemodule multiple times
in your interpreter is *not* reloading the module. This is because the first step in
the resolution protocol import is to see if the desired module is already in the
sys.modules dictionary and then to *not* import it if it already there.

Sometimes you just need a specific function from a given module. The from
<module> import <name> syntax handles that case,

```
>>> # will overwrite existing definitions in the namespace!
>>> from math import sqrt
>>> sqrt(a) # shorter to use
3.8078865529319543
```

Note that you have to use the importlib.reload function to re-import modules
into the workspace. Importantly, importlib.reload will not work with the
from syntax used above. Thus, if you are developing code and constantly reloading
it, it is better to keep the top level module name so you can keep reloading it with
importlib.reload.

There are many places to hijack the import process. These make it possible to
create virtual environments and to customize execution, but it is best to stick with
the well-developed solutions for these cases (see conda below).

3.2   Writing and Using Your Own Modules

Beyond using the amazingly excellent standard library, you may want to share your
codes with your colleagues. The easiest way to do this is to put your code in a file
and send it to them. When interactively developing code to be distributed this way,

you must understand how your module is updated (or not) in the active interpreter. For example, put the following code in a separate file named `mystuff.py`:

```python
def sqrt(x):
    return x*x
```

and then return to the interactive session

```python
>>> from mystuff import sqrt
>>> sqrt(3)
9
>>> import mystuff
>>> dir(mystuff)
['__builtins__', '__cached__', '__doc__',
↪    '__file__','__loader__','__name__', '__package__',
↪    '__spec__', 'sqrt']
>>> mystuff.sqrt(3)
9
```

Now, add the following function to your `mystuff.py` file:

```python
def poly_func(x):
    return 1+x+x*x
```

And change your previous function in the file:

```python
def sqrt(x):
    return x/2.
```

and then return to the interactive session

```python
mystuff.sqrt(3)
```

Did you get what you expected?

```python
mystuff.poly_func(3)
```

You have to `importlib.reload` in order to get new changes in your file into the interpreter. A directory called __pycache__ will automatically appear in the same directory as `mystuff.py`. This is where Python stores the compiled byte-codes for the module so that Python does not have to re-compile it from scratch every time you import `mystuff.py`. This directory will automatically be refreshed every time you make changes to `mystuff.py`. It is important to *never* include the __pycache__ directory into your Git repository because when others clone your repository, if the filesystem gets the timestamps wrong, it could be that __pycache__ falls out of sync with the source code. This is a painful bug because others may make changes to the `mystuff.py` file and those changes will not be implemented when `mystuff` module is imported because Python is still using the version in __pycache__. If you are working with Python 2.x, then the compiled Python byte-codes are stored in the same directory without __pycache__ as `.pyc` files. These should also never be included in your Git repository for the same reason.

> **Programming Tip: IPython Automated Reloading**
> IPython provides some autoreloading via the `%autoreload` magic, but it comes with lots of caveats. Better to explicitly use `importlib.reload`.

3.2.1   Using a Directory as a Module

Beyond sticking all your Python code into a single file, you can use a directory to organize your code into separate files. The trick is to put a `__init__.py` file in the top level of the directory you want to import from. The file can be empty. For example,

```
package/
    __init__.py
    moduleA.py
```

So, if `package` is in your path, you can do `import package`. If `__init__.py` is empty, then this does nothing. To get any of the code in `moduleA.py`, you have to explicitly import it as `import package.moduleA`, and then you can get the contents of that module. For example,

```
package.moduleA.foo()
```

runs the `foo` function in the `moduleA.py` file. If you want to make `foo` available upon importing `package`, then you have to put `from .moduleA import foo` in the `__init__.py` file. Relative import notation is required in Python 3. Then, you can do `import package` and then run the function as `package.foo()`. You can also do `from package import foo` to get `foo` directly. When developing your own modules, you can have fine-grained control of which packages are imported using relative imports.

3.3   Dynamic Importing

In case you do not know the names of the modules that you need to import ahead of time, the `__import__` function can load modules from a specified list of module names.

```
>>> sys = __import__('sys')  # import module from string argument
>>> sys.version
'3.8.3 (default, May 19 2020, 18:47:26) \n[GCC 7.3.0]'
```

Programming Tip: Using `__main__`
Namespaces distinguish between importing and running a Python script.

```
if __name__ == '__main__':
    # these statements are not executed during import
    # do run statements here
```

There is also `__file__` which is the filename of the imported file. In addition to the `__init__.py` file, you can put a `__main__.py` file in the top level of your module directory if you want to call your module using the `-m` switch on the commandline. Doing this means the `__init__.py` function will also run.

3.4   Getting Modules from the Web

Python packaging has really improved over the last few years. This had always been a sore point, but now it is a *lot* easier to deploy and maintain Python codes that rely on linked libraries across multiple platforms. Python packages in the main Python Package Index support `pip`.

```
% pip install name_of_module
```

This figures out all the dependencies and installs them, too. There are many flags that control how and where packages are installed. You do not need root access to use this. See the `--user` flag for non-root access. Modern Python packaging relies on the so-called wheel files that include fragments of compiled libraries that the module depends on. This usually works, but if you have problems on the Windows platform, there is a treasure trove of modules for Windows in the form of wheel files at Christoph Gohlke's lab at UCI.[1]

3.5   Conda Package Management

You should really use `conda` whenever possible. It soothes so many package management headaches, and you do not need administrator/root rights to use it effectively. The *anaconda* toolset is a curated list of scientific packages that is supported by the Anaconda company. This has just about all the scientific packages you want. Outside of this support, the community also supports a longer list of scientific packages as `conda-forge`. You can add `conda-forge` to your usual repository list with the following:

---

[1] See https://www.l.uci.edu/~gohlke/pythonlibs..

```
Terminal> conda config --add channels conda-forge
```

And then you can install new modules as in the following:

```
Terminal> conda install name_of_module
```

Additionally, `conda` also facilitates self-contained sub-environments that are a great way to safely experiment with codes and even different Python versions. This can be key for automated provisioning of virtual machines in a cloud-computing environment. For example, the following will create an environment named `my_test_env` with the Python version 3.7.

```
Terminal> conda create -n my_test_env python=3.7
```

The difference between `pip` and `conda` is that `pip` will use the requirements for the desired package to ensure that to install any missing modules. The `conda` package manager will do the same but will additionally determine if there are any conflicts in the versions of the desired packages and its dependencies versus the existing installation and provide a warning beforehand.[2] This avoids the problem of overwriting the dependencies of a pre-existing package to satisfy a new installation. The best practice is to prefer `conda` when dealing with scientific codes with many linked libraries and then rely on `pip` for the pure Python codes. Sometimes it is necessary to use both because certain desired Python modules might not yet be supported by `conda`. This can become a problem when `conda` does not know how to integrate the new `pip` installed packages for internal management. The `conda` documentation has more information and bear in mind that `conda` is also under constant development.

Another way to create virtual environments is with the `venv` (or `virtualenv`), which comes with Python itself. Again, this is a good idea for `pip` installed packages and particularly for pure Python codes, but `conda` is a better alternative for scientific programs. Nonetheless, virtual environments created with `venv` or `virtualenv` are particularly useful for segregating commandline programs that may have odd dependencies that you do not want to carry around or have interfere with other installations.

Programming Tip: Mamba Package Manager
The `mamba` package manager is multiple times faster than `conda` because of its more efficient satisfiability solver, and is an excellent replacement for `conda`.

The following are listed in the Bibliography as excellent books for learning more about Python: [1–4, 4–13, 13–15].

---

[2]To resolve conflicts, `conda` implements a satisfiability solver (SAT), which is a classic combinatorial problem.

References

1. D.M. Beazley, *Python Essential Reference* (Addison-Wesley, Boston, 2009)
2. D. Beazley, B.K. Jones, *Python Cookbook: Recipes for Mastering Python 3* (O'Reilly Media, Newton, 2013)
3. N. Ceder, *The Quick Python Book*. (Manning Publications, Shelter Island, 2018)
4. D. Hellmann, *The Python 3 Standard Library by Example* Developer's Library (Pearson Education, London, 2017)
5. C. Hill, *Learning Scientific Programming With Python* (Cambridge University Press, Cambridge, 2020)
6. D. Kuhlman, *A Python Book: Beginning Python, Advanced Python, and Python Exercises* (Platypus Global Media, Washington, 2011)
7. H.P. Langtangen, *A Primer on Scientific Programming With Python*. Texts in Computational Science and Engineering (Springer, Berlin, Heidelberg, 2016)
8. M. Lutz, *Learning Python: Powerful Object-Oriented Programming*. Safari Books Online (O'Reilly Media, Newton, 2013)
9. M. Pilgrim, *Dive Into Python 3*. Books for Professionals by Professionals (Apress, New York, 2010)
10. K. Reitz, T. Schlusser, *The Hitchhiker's Guide to Python: Best Practices for Development* (O'Reilly Media, Newton, 2016)
11. C. Rossant, *Learning IPython for Interactive Computing and Data Visualization* (Packt Publishing, Birmingham, 2015)
12. Z.A. Shaw, *Learn Python the Hard Way: Release 2.0*. Lulu.com (2012)
13. M. Summerfield, Python in Practice: Create Better Programs Using Concurrency, Libraries, and Patterns (Pearson Education, London, 2013)
14. J. Unpingco, *Python for Signal Processing: Featuring IPython Notebooks* (Springer International Publishing, Cham, 2016)
15. J. Unpingco, *Python for Probability, Statistics, and Machine Learning*, 2nd edn. (Springer International Publishing, Cham, 2019)

Chapter 4
Numpy

Numpy provides a unified way to manage numerical arrays in Python. It consolidated the best ideas from many prior approaches to numerical arrays in Python. It is the cornerstone of many other scientific computing Python modules. To understand and to be effective with scientific computing and Python, a solid grasp on Numpy is essential!

```
>>> import numpy as np # naming convention
```

4.1 Dtypes

Although Python is dynamically typed, Numpy permits fine-grained specification of number types using dtypes,

```
>>> a = np.array([0],np.int16) # 16-bit integer
>>> a.itemsize # in 8-bit bytes
2
>>> a.nbytes
2
>>> a = np.array([0],np.int64) # 64-bit integer
>>> a.itemsize
8
```

Numerical arrays follow the same pattern,

```
>>> a = np.array([0,1,23,4],np.int64) # 64-bit integer
>>> a.shape
(4,)
>>> a.nbytes
32
```

Note that you *cannot* tack on extra elements to a Numpy array after creation,

```
>>> a = np.array([1,2])
>>> a[2] = 32
```

© The Editor(s) (if applicable) and The Author(s), under exclusive license
to Springer Nature Switzerland AG 2021
J. Unpingco, *Python Programming for Data Analysis*,
https://doi.org/10.1007/978-3-030-68952-0_4

```
Traceback (most recent call last):
  File "<stdin>", line 1, in <module>
IndexError: index 2 is out of bounds for axis 0 with size 2
```

This is because the block of memory has already been delineated and Numpy will not allocate new memory and copy data without explicit instruction. Also, once you create the array with a specific dtype, assigning to that array will cast to that type. For example,

```
>>> x = np.array(range(5), dtype=int)
>>> x[0] = 1.33 # float assignment does not match dtype=int
>>> x
array([1, 1, 2, 3, 4])
>>> x[0] = 'this is a string'
Traceback (most recent call last):
  File "<stdin>", line 1, in <module>
ValueError: invalid literal for int() with base 10: 'this is a
↪   string'
```

This is different from systems like Matlab because the copy/view array semantics are so fundamentally different.

4.2 Multidimensional Arrays

Multidimensional arrays follow the same pattern,

```
>>> a = np.array([[1,3],[4,5]]) # omitting dtype picks default
>>> a
array([[1, 3],
       [4, 5]])
>>> a.dtype
dtype('int64')
>>> a.shape
(2, 2)
>>> a.nbytes
32
>>> a.flatten()
array([1, 3, 4, 5])
```

The maximum limit on the number of dimensions depends on how it is configured during the Numpy build (usually thirty-two). Numpy offers many ways to build arrays automatically,

```
>>> a = np.arange(10) # analogous to range()
>>> a
array([0, 1, 2, 3, 4, 5, 6, 7, 8, 9])
>>> a = np.ones((2,2))
>>> a
array([[1., 1.],
       [1., 1.]])
>>> a = np.linspace(0,1,5)
>>> a
```

```
array([0.  , 0.25, 0.5 , 0.75, 1.  ])
>>> X,Y = np.meshgrid([1,2,3],[5,6])
>>> X
array([[1, 2, 3],
       [1, 2, 3]])
>>> Y
array([[5, 5, 5],
       [6, 6, 6]])
>>> a = np.zeros((2,2))
>>> a
array([[0., 0.],
       [0., 0.]])
```

You can also create Numpy arrays using functions,

```
>>> np.fromfunction(lambda i,j: abs(i-j)<=1, (4,4))
array([[ True,  True, False, False],
       [ True,  True,  True, False],
       [False,  True,  True,  True],
       [False, False,  True,  True]])
```

Numpy arrays can also have fieldnames,

```
>>> a = np.zeros((2,2), dtype=[('x','f4')])
>>> a['x']
array([[0., 0.],
       [0., 0.]], dtype=float32)
>>> x = np.array([(1,2)], dtype=[('value','f4'),
...                              ('amount','c8')])
>>> x['value']
array([1.], dtype=float32)
>>> x['amount']
array([2.+0.j], dtype=complex64)
>>> x = np.array([(1,9),(2,10),(3,11),(4,14)],
...              dtype=[('value','f4'),
...                     ('amount','c8')])
>>> x['value']
array([1., 2., 3., 4.], dtype=float32)
>>> x['amount']
array([ 9.+0.j, 10.+0.j, 11.+0.j, 14.+0.j], dtype=complex64)
```

Numpy arrays can also be accessed by their attributes using `recarray`,

```
>>> y = x.view(np.recarray)
>>> y.amount # access as attribute
array([ 9.+0.j, 10.+0.j, 11.+0.j, 14.+0.j], dtype=complex64)
>>> y.value # access as attribute
array([1., 2., 3., 4.], dtype=float32)
```

4.3   Reshaping and Stacking Numpy Arrays

Arrays can be stacked horizontally and vertically,

```
>>> x = np.arange(5)
>>> y = np.array([9,10,11,12,13])
```

```
>>> np.hstack([x,y]) # stack horizontally
array([ 0,   1,   2,   3,   4,   9, 10, 11, 12, 13])
>>> np.vstack([x,y]) # stack vertically
array([[ 0,   1,   2,   3,   4],
       [ 9, 10, 11, 12, 13]])
```

There is also a dstack method if you want to stack in the third *depth* dimension. Numpy np.concatenate handles the general arbitrary-dimension case. In some codes (e.g., scikit-learn), you may find the terse np.c_ and np.r_ used to stack arrays column-wise and row-wise:

```
>>> np.c_[x,x] # column-wise
array([[0,   0],
       [1,   1],
       [2,   2],
       [3,   3],
       [4,   4]])
>>> np.r_[x,x] # row-wise
array([0, 1, 2, 3, 4, 0, 1, 2, 3, 4])
```

4.4 Duplicating Numpy Arrays

Numpy has a repeat function for duplicating elements and a more generalized version in tile that lays out a block matrix of the specified shape,

```
>>> x=np.arange(4)
>>> np.repeat(x,2)
array([0, 0, 1, 1, 2, 2, 3, 3])
>>> np.tile(x,(2,1))
array([[0, 1, 2, 3],
       [0, 1, 2, 3]])
>>> np.tile(x,(2,2))
array([[0, 1, 2, 3, 0, 1, 2, 3],
       [0, 1, 2, 3, 0, 1, 2, 3]])
```

You can also have non-numerics like strings as items in the array

```
>>> np.array(['a','b','cow','deep'])
array(['a', 'b', 'cow', 'deep'], dtype='<U4')
```

Note that the 'U4' refers to string of length 4, which is the longest string in the sequence.

Reshaping Numpy arrays Numpy arrays can be reshaped after creation,

```
>>> a = np.arange(10).reshape(2,5)
>>> a
array([[0, 1, 2, 3, 4],
       [5, 6, 7, 8, 9]])
```

For the truly lazy, you can replace one of the dimensions above by negative one (i.e., reshape(-1,5)), and Numpy will figure out the conforming other dimension. The array transpose method operation is the same as the .T attribute,

```
>>> a.transpose()
array([[0,  5],
       [1,  6],
       [2,  7],
       [3,  8],
       [4,  9]])
>>> a.T
array([[0,  5],
       [1,  6],
       [2,  7],
       [3,  8],
       [4,  9]])
```

The conjugate transpose (i.e., Hermitian transpose) is the .H attribute.

4.5   Slicing, Logical Array Operations

Numpy arrays follow the same zero-indexed slicing logic as Python lists and strings:

```
>>> x = np.arange(50).reshape(5,10)
>>> x
array([[ 0,  1,  2,  3,  4,  5,  6,  7,  8,  9],
       [10, 11, 12, 13, 14, 15, 16, 17, 18, 19],
       [20, 21, 22, 23, 24, 25, 26, 27, 28, 29],
       [30, 31, 32, 33, 34, 35, 36, 37, 38, 39],
       [40, 41, 42, 43, 44, 45, 46, 47, 48, 49]])
```

The colon character means take all along the indicated dimension

```
>>> x[:,0]
array([ 0, 10, 20, 30, 40])
>>> x[0,:]
array([0, 1, 2, 3, 4, 5, 6, 7, 8, 9])
>>> x = np.arange(50).reshape(5,10) # reshaping arrays
>>> x
array([[ 0,  1,  2,  3,  4,  5,  6,  7,  8,  9],
       [10, 11, 12, 13, 14, 15, 16, 17, 18, 19],
       [20, 21, 22, 23, 24, 25, 26, 27, 28, 29],
       [30, 31, 32, 33, 34, 35, 36, 37, 38, 39],
       [40, 41, 42, 43, 44, 45, 46, 47, 48, 49]])
>>> x[:,0] # any row, 0th column
array([ 0, 10, 20, 30, 40])
>>> x[0,:] # any column, 0th row
array([0, 1, 2, 3, 4, 5, 6, 7, 8, 9])
>>> x[1:3,4:6]
array([[14, 15],
       [24, 25]])
>>> x = np.arange(2*3*4).reshape(2,3,4) # reshaping arrays
>>> x
array([[[ 0,  1,  2,  3],
        [ 4,  5,  6,  7],
        [ 8,  9, 10, 11]],
```

```
          [[12,  13,  14,  15],
           [16,  17,  18,  19],
           [20,  21,  22,  23]]]])
>>> x[:,1,[2,1]] # index each dimension
array([[ 6,   5],
       [18,  17]])
```

Numpy's where function can find array elements according to specific logical
criteria. Note that np.where returns a tuple of Numpy indices,

```
>>> np.where(x % 2 == 0)
(array([0, 0, 0, 0, 0, 0, 1, 1, 1, 1, 1, 1]),
array([0, 0, 1, 1, 2, 2, 0, 0, 1, 1, 2, 2]),
array([0, 2, 0, 2, 0, 2, 0, 2, 0, 2, 0, 2]))
>>> x[np.where(x % 2 == 0)]
array([ 0,   2,   4,   6,   8, 10, 12, 14, 16, 18, 20, 22])
>>> x[np.where(np.logical_and(x % 2 == 0,x < 9))] # also
↪  logical_or, etc.
array([0, 2, 4, 6, 8])
```

Additionally, Numpy arrays can be indexed by logical Numpy arrays where only
the corresponding True entries are selected,

```
>>> a = np.arange(9).reshape((3,3))
>>> a
array([[0, 1, 2],
       [3, 4, 5],
       [6, 7, 8]])
>>> b = np.fromfunction(lambda i,j: abs(i-j) <= 1, (3,3))
>>> b
array([[ True,   True, False],
       [ True,   True,  True],
       [False,   True,  True]])
>>> a[b]
array([0, 1, 3, 4, 5, 7, 8])
>>> b = (a>4)
>>> b
array([[False, False, False],
       [False, False,  True],
       [ True,  True,  True]])
>>> a[b]
array([5, 6, 7, 8])
```

4.6 Numpy Arrays and Memory

Numpy uses pass-by-reference semantics so that slice operations are *views* into
the array without implicit copying, which is consistent with Python's semantics.
This is particularly helpful with large arrays that already strain available memory.
In Numpy terminology, *slicing* creates views (no copying) and advanced indexing
creates copies. Let us start with advanced indexing.

   If the indexing object (i.e., the item between the brackets) is a non-tuple sequence
object, another Numpy array (of type integer or boolean), or a tuple with at least

one sequence object or Numpy array, then indexing creates copies. For the above example, to extend and copy an existing array in Numpy, you have to do something like the following:

```
>>> x = np.ones((3,3))
>>> x
array([[1., 1., 1.],
       [1., 1., 1.],
       [1., 1., 1.]])
>>> x[:,[0,1,2,2]] # notice duplicated last dimension
array([[1., 1., 1., 1.],
       [1., 1., 1., 1.],
       [1., 1., 1., 1.]])
>>> y=x[:,[0,1,2,2]] # same as above, but do assign it to y
```

Because of advanced indexing, the variable y has its own memory because the relevant parts of x were copied. To prove it, we assign a new element to x and see that y is not updated:

```
>>> x[0,0]=999 # change element in x
>>> x # changed
array([[999.,    1.,    1.],
       [  1.,    1.,    1.],
       [  1.,    1.,    1.]])
>>> y # not changed!
array([[1., 1., 1., 1.],
       [1., 1., 1., 1.],
       [1., 1., 1., 1.]])
```

However, if we start over and construct y by slicing (which makes it a view) as shown below, then the change we made *does* affect y because a view is just a window into the same memory:

```
>>> x = np.ones((3,3))
>>> y = x[:2,:2] # view of upper left piece
>>> x[0,0] = 999 # change value
>>> x   # see the change?
array([[999.,    1.,    1.],
       [  1.,    1.,    1.],
       [  1.,    1.,    1.]])
>>> y
array([[999.,    1.],
       [  1.,    1.]])
```

Note that if you want to explicitly force a copy without any indexing tricks, you can do y=x.copy(). The code below works through another example of advanced indexing versus slicing:

```
>>> x = np.arange(5) # create array
>>> x
array([0, 1, 2, 3, 4])
>>> y=x[[0,1,2]] # index by integer list to force copy
>>> y
array([0, 1, 2])
>>> z=x[:3]      # slice creates view
```

```
>>> z                    # note y and z have same entries
array([0, 1, 2])
>>> x[0]=999             # change element of x
>>> x
array([999,    1,    2,    3,    4])
>>> y                    # note y is unaffected,
array([0, 1, 2])
>>> z                    # but z is (it's a view).
array([999,    1,    2])
```

In this example, y is a copy, not a view, because it was created using advanced indexing, whereas z was created using slicing. Thus, even though y and z have the same entries, only z is affected by changes to x. Note that the flags.ownsdata property of Numpy arrays can help sort this out until you get used to it.

Overlapping Numpy Arrays Manipulating memory using views is particularly powerful for signal and image processing algorithms that require overlapping fragments of memory. The following is an example of how to use advanced Numpy to create overlapping blocks that do not actually consume additional memory,

```
>>> from numpy.lib.stride_tricks import as_strided
>>> x = np.arange(16).astype(np.int32)
>>> y=as_strided(x,(7,4),(8,4)) # overlapped entries
>>> y
array([[ 0,    1,    2,    3],
       [ 2,    3,    4,    5],
       [ 4,    5,    6,    7],
       [ 6,    7,    8,    9],
       [ 8,    9,   10,   11],
       [10,   11,   12,   13],
       [12,   13,   14,   15]], dtype=int32)
```

The above code creates a range of integers and then overlaps the entries to create a 7x4 Numpy array. The final argument in the as_strided function are the strides, which are the steps in bytes to move in the row and column dimensions, respectively. Thus, the resulting array steps four bytes in the column dimension and eight bytes in the row dimension. Because the integer elements in the Numpy array are four bytes, this is equivalent to moving by one element in the column dimension and by two elements in the row dimension. The second row in the Numpy array starts at eight bytes (two elements) from the first entry (i.e., 2) and then proceeds by four bytes (by one element) in the column dimension (i.e., 2,3,4,5). The important part is that memory is re-used in the resulting 7x4 Numpy array. The code below demonstrates this by reassigning elements in the original x array. The changes show up in the y array because they point at the same allocated memory:

```
>>> x[::2] = 99 # assign every other value
>>> x
array([99, 1, 99, 3, 99, 5, 99, 7, 99, 9, 99, 11, 99, 13, 99,
 ↪  15],
      dtype=int32)
>>> y # the changes appear because y is a view
array([[99,    1, 99,    3],
```

```
    [99,   3, 99,   5],
    [99,   5, 99,   7],
    [99,   7, 99,   9],
    [99,   9, 99,  11],
    [99,  11, 99,  13],
    [99,  13, 99,  15]], dtype=int32)
```

Bear in mind that `as_strided` does *not* check that you stay within memory block bounds. So, if the size of the target matrix is not filled by the available data, the remaining elements will come from whatever bytes are at that memory location. In other words, there is no default filling by zeros or other strategy that defends memory block bounds. One defense is to explicitly control the dimensions as in the following code:

```
>>> n = 8 # number of elements
>>> x = np.arange(n) # create array
>>> k = 5 # desired number of rows
>>> y = as_strided(x,(k,n-k+1),(x.itemsize,)*2)
>>> y
array([[0, 1, 2, 3],
       [1, 2, 3, 4],
       [2, 3, 4, 5],
       [3, 4, 5, 6],
       [4, 5, 6, 7]])
```

4.7   Numpy Memory Data Structures

Let us examine the following data structure `typedef` in the Numpy source code:

```
typedef struct PyArrayObject {
    PyObject_HEAD

        /* Block of memory */
        char *data;

    /* Data type descriptor */
    PyArray_Descr *descr;

    /* Indexing scheme */
    int nd;
    npy_intp *dimensions;
    npy_intp *strides;

    /* Other stuff */
    PyObject *base;
    int flags;
    PyObject *weakreflist;
} PyArrayObject;0
```

Let us create a Numpy array of 16-bit integers and explore it:

```
>>> x = np.array([1], dtype=np.int16)
```

We can see the raw data using the `x.data` attribute,

```
>>> bytes(x.data)
b'\x01\x00'
```

Notice the orientation of the bytes. Now, change the `dtype` to unsigned two-byte big-endian integer,

```
>>> x = np.array([1], dtype='>u2')
>>> bytes(x.data)
b'\x00\x01'
```

Again notice the orientation of the bytes. This is what little/big endian means for data in memory. We can create Numpy arrays from bytes directly using `frombuffer`, as in the following. Note the effect of using different `dtypes`,

```
>>> np.frombuffer(b'\x00\x01',dtype=np.int8)
array([0, 1], dtype=int8)
>>> np.frombuffer(b'\x00\x01',dtype=np.int16)
array([256], dtype=int16)
>>> np.frombuffer(b'\x00\x01',dtype='>u2')
array([1], dtype=uint16)
```

Once a `ndarray` is created, you can re-cast it to a different dtype or change the dtype of the `view`. Roughly speaking, casting copies data. For example,

```
>>> x = np.frombuffer(b'\x00\x01',dtype=np.int8)
>>> x
array([0, 1], dtype=int8)
>>> y = x.astype(np.int16)
>>> y
array([0, 1], dtype=int16)
>>> y.flags['OWNDATA'] # y is a copy
True
```

Alternatively, we can re-interpret the data using a `view`,

```
>>> y = x.view(np.int16)
>>> y
array([256], dtype=int16)
>>> y.flags['OWNDATA']
False
```

Note that `y` is not new memory, it is just referencing the existing memory and re-interpreting it using a different `dtype`.

Numpy Memory Strides The strides of the typedef above concern how Numpy moves between and across arrays. A *stride* is the number of bytes to reach the next consecutive array element. There is one stride per dimension. Consider the following Numpy array:

```
>>> x = np.array([[1, 2, 3], [4, 5, 6], [7, 8, 9]], dtype=np.int8)
>>> bytes(x.data)
b'\x01\x02\x03\x04\x05\x06\x07\x08\t'
>>> x.strides
(3, 1)
```

Thus, if we want to index x[1,2], we have to use the following offset:

```
>>> offset = 3*1+1*2
>>> x.flat[offset]
6
```

Numpy supports C-order (i.e., column-wise) and Fortran-order (i.e., row-wise). For example,

```
>>> x = np.array([[1, 2, 3], [7, 8, 9]], dtype=np.int8,order='C')
>>> x.strides
(3, 1)
>>> x = np.array([[1, 2, 3], [7, 8, 9]], dtype=np.int8,order='F')
>>> x.strides
(1, 2)
```

Note the difference between the strides for the two orderings. For C-order, it takes 3 bytes to move between the rows and 1 byte to move between columns, whereas for Fortran-order, it takes 1 byte to move between rows, but 2 bytes to move between columns. This pattern continues for higher dimensions:

```
>>> x = np.arange(125).reshape((5,5,5)).astype(np.int8)
>>> x.strides
(25, 5, 1)
>>> x[1,2,3]
38
```

To get the [1,2,3] element using byte offsets, we can do the following:

```
>>> offset = (25*1 + 5*2 +1*3)
>>> x.flat[offset]
38
```

Once again, creating views by slicing just changes the stride!

```
>>> x = np.arange(3,dtype=np.int32)
>>> x.strides
(4,)
>>> y = x[::-1]
>>> y.strides
(-4,)
```

Transposing also just swaps the stride,

```
>>> x = np.array([[1, 2, 3], [7, 8, 9]], dtype=np.int8,order='F')
>>> x.strides
(1, 2)
>>> y = x.T
>>> y.strides # negative!
(2, 1)
```

In general, reshaping does *not* just alter the stride but may sometimes make copies of the data. The memory layout (i.e., strides) can affect performance because of the CPU cache. The CPU pulls in data from main memory in blocks so that if many items can consecutively be operated on in a single block then that reduces the number of transfers needed from main memory that speeds up computation.

4.8   Array Element-Wise Operations

The usual pairwise arithmetic operations are element-wise in Numpy:

```
>>> x*3
array([[ 3,   6,   9],
       [21,  24,  27]], dtype=int8)
>>> y = x/x.max()
>>> y
array([[0.11111111, 0.22222222, 0.33333333],
       [0.77777778, 0.88888889, 1.        ]])
>>> np.sin(y) * np.exp(-y)
array([[0.09922214, 0.17648072, 0.23444524],
       [0.32237812, 0.31917604, 0.30955988]])
```

> **Programming Tip: Beware Numpy in-place Operations**
> It is easy to mess up in-place operations such as x -= x.T for Numpy arrays
> so these should be avoided in general and can lead to hard-to-find bugs later.

4.9   Universal Functions

Now that we know how to create and manipulate Numpy arrays, let us consider how
to compute with other Numpy features. Universal functions (*ufuncs*) are Numpy
functions that are optimized to compute Numpy arrays at the C-level (i.e., outside
the Python interpreter). Let us compute the trigonometric sine:

```
>>> a = np.linspace(0,1,20)
>>> np.sin(a)
array([0.        , 0.05260728, 0.10506887, 0.15723948, 0.20897462,
       0.26013102, 0.310567  , 0.36014289, 0.40872137, 0.45616793,
       0.50235115, 0.54714315, 0.59041986, 0.63206143, 0.67195255,
       0.70998273, 0.74604665, 0.78004444, 0.81188195, 0.84147098])
```

Note that Python has a built-in math module with its own sine function:

```
>>> from math import sin
>>> [sin(i) for i in a]
[0.0, 0.05260728333807213, 0.10506887376594912,
0.15723948186175024, 0.20897462406278547, 0.2601310228046501,
0.3105670033203749, 0.3601428860007191, 0.40872137322898616,
0.4561679296190457, 0.5023511546035125, 0.547143146340223,
0.5904198559291864, 0.6320614309590333, 0.6719525474315213,
0.7099827291448582, 0.7460466536513234, 0.7800444439418607,
0.811819450498316, 0.8414709848078965]
```

The output is a list, not a Numpy array, and to process all the elements of a, we
had to use the list comprehensions to compute the sine. This is because Python's

`math` function only works one at a time with each member of the array. Numpy's sine function does not need these extra semantics because the computing runs in the Numpy C-code *outside* of the Python interpreter. This is where Numpy's 200–300 times speed-up over plain Python code comes from. So, doing the following:

```
>>> np.array([sin(i) for i in a])
array([0.        , 0.05260728, 0.10506887, 0.15723948, 0.20897462,
       0.26013102, 0.310567  , 0.36014289, 0.40872137, 0.45616793,
       0.50235115, 0.54714315, 0.59041986, 0.63206143, 0.67195255,
       0.70998273, 0.74604665, 0.78004444, 0.81188195, 0.84147098])
```

entirely defeats the purpose of using Numpy. Always use Numpy ufuncs whenever possible!

```
>>> np.sin(a)
array([0.        , 0.05260728, 0.10506887, 0.15723948, 0.20897462,
       0.26013102, 0.310567  , 0.36014289, 0.40872137, 0.45616793,
       0.50235115, 0.54714315, 0.59041986, 0.63206143, 0.67195255,
       0.70998273, 0.74604665, 0.78004444, 0.81188195, 0.84147098])
```

4.10   Numpy Data Input/Output

Numpy makes it easy to move data in and out of files:

```
>>> x = np.loadtxt('sample1.txt')
>>> x
array([[ 0.,  0.],
       [ 1.,  1.],
       [ 2.,  4.],
       [ 3.,  9.],
       [ 4., 16.],
       [ 5., 25.],
       [ 6., 36.],
       [ 7., 49.],
       [ 8., 64.],
       [ 9., 81.]])
>>> # each column with different type
>>> x = np.loadtxt('sample1.txt',dtype='f4,i4')
>>> x
array([(0.,  0), (1.,  1), (2.,  4), (3.,  9), (4., 16), (5.,
↪   25),
       (6., 36), (7., 49), (8., 64), (9., 81)],
      dtype=[('f0', '<f4'), ('f1', '<i4')])
```

Numpy arrays can be saved with the corresponding `np.savetxt` function.

4.11   Linear Algebra

Numpy has direct access to the proven LAPACK/BLAS linear algebra code. The main entryway to linear algebra functions in Numpy is via the `linalg` submodule,

```
>>> np.linalg.eig(np.eye(3)) # runs underlying LAPACK/BLAS
(array([1., 1., 1.]), array([[1., 0., 0.],
       [0., 1., 0.],
       [0., 0., 1.]]))
>>> np.eye(3)*np.arange(3) # does this work as expected?
array([[0., 0., 0.],
       [0., 1., 0.],
       [0., 0., 2.]])
```

To get matrix row–column products, you can use the `matrix` object,

```
>>> np.eye(3)*np.matrix(np.arange(3)).T # row-column multiply,
matrix([[0.],
        [1.],
        [2.]])
```

More generally, you can use Numpy's `dot` product,

```
>>> a = np.eye(3)
>>> b = np.arange(3).T
>>> a.dot(b)
array([0., 1., 2.])
>>> b.dot(b)
5
```

The advantage of `dot` is that it works in arbitrary dimensions. This is handy for tensor-like contractions (see Numpy `tensordot` for more info). Since Python 3.6, there is also the `@` notation for Numpy matrix multiplication

```
>>> a = np.eye(3)
>>> b = np.arange(3).T
>>> a @ b
array([0., 1., 2.])
```

4.12   Broadcasting

Broadcasting is incredibly powerful, but it takes time to understand. Quoting from the *Guide to NumPy* by Travis Oliphant [1]:

1. All input arrays with ndim smaller than the input array of largest ndim have 1's pre-pended to their shapes.
2. The size in each dimension of the output shape is the maximum of all the input shapes in that dimension.
3. An input can be used in the calculation if it is the shape in a particular dimension either matches the output shape or has value exactly 1.
4. If an input has a dimension size of 1 in its shape, the first data entry in that dimension will be used for all calculations along that dimension. In other words, the stepping machinery of the ufunc will simply not step along that dimension when otherwise needed (the stride will be 0 for that dimension).

An easier way to think about these rules is the following:

1. If the array shapes have different lengths, then left-pad the smaller shape with ones.
2. If any corresponding dimension does not match, make copies along the 1-dimension.
3. If any corresponding dimension does not have a one in it, raise an error.

Some examples will help. Consider these two arrays:

```
>>> x = np.arange(3)
>>> y = np.arange(5)
```

and you want to compute the element-wise product of these. The problem is that this operation is not defined for arrays of different shapes. We can define what this element-wise product means in this case with the following loop,

```
>>> out = []
>>> for i in x:
...     for j in y:
...         out.append(i*j)
...
>>> out
[0, 0, 0, 0, 0, 0, 1, 2, 3, 4, 0, 2, 4, 6, 8]
```

But now we have lost the input dimensions of x and y. We can conserve these by reshaping the output as follows:

```
>>> out=np.array(out).reshape(len(x),-1) # -1 means infer the
↪   remaining dimension
>>> out
array([[0, 0, 0, 0, 0],
       [0, 1, 2, 3, 4],
       [0, 2, 4, 6, 8]])
```

Another way to think about what we have just computed is as the matrix outer product,

```
>>> from numpy import matrix
>>> out=matrix(x).T * y
>>> out
matrix([[0, 0, 0, 0, 0],
        [0, 1, 2, 3, 4],
        [0, 2, 4, 6, 8]])
```

But how can you generalize this to handle multiple dimensions? Let us consider adding a *singleton* dimension to y as

```
>>> x[:,None].shape
(3, 1)
```

We can use np.newaxis instead of None for readability. Now, if we try this directly, broadcasting will handle the incompatible dimensions by making copies along the singleton dimension:

```
>>> x[:,None]*y
array([[0,  0,  0,  0,  0],
       [0,  1,  2,  3,  4],
       [0,  2,  4,  6,  8]])
```

and this works with more complicated expressions:

```
>>> from numpy import cos
>>> x[:,None]*y + cos(x[:,None]+y)
array([[ 1.        ,  0.54030231,  -0.41614684,  -0.9899925 ,  -0.65364362],
       [ 0.54030231,  0.58385316,   1.0100075 ,   2.34635638,  4.28366219],
       [-0.41614684,  1.0100075 ,   3.34635638,   6.28366219,  8.96017029]])
```

But what if you do not like the shape of the resulting array?

```
>>> x*y[:,None] # change the placement of the singleton dimension
array([[0,  0,  0],
       [0,  1,  2],
       [0,  2,  4],
       [0,  3,  6],
       [0,  4,  8]])
```

Now, let us consider a bigger example

```
>>> X = np.arange(2*4).reshape(2,4)
>>> Y = np.arange(3*5).reshape(3,5)
```

where you want to element-wise multiply these two together. The result will be a 2 x 4 x 3 x 5 multidimensional matrix:

```
>>> X[:,:,None,None] * Y
array([[[[ 0,  0,  0,  0,  0],
         [ 0,  0,  0,  0,  0],
         [ 0,  0,  0,  0,  0]],

        [[ 0,  1,  2,  3,  4],
         [ 5,  6,  7,  8,  9],
         [10, 11, 12, 13, 14]],

        [[ 0,  2,  4,  6,  8],
         [10, 12, 14, 16, 18],
         [20, 22, 24, 26, 28]],

        [[ 0,  3,  6,  9, 12],
         [15, 18, 21, 24, 27],
         [30, 33, 36, 39, 42]]],


       [[[ 0,  4,  8, 12, 16],
         [20, 24, 28, 32, 36],
         [40, 44, 48, 52, 56]],

        [[ 0,  5, 10, 15, 20],
         [25, 30, 35, 40, 45],
```

```
        [50, 55, 60, 65, 70]],

       [[ 0,  6, 12, 18, 24],
        [30, 36, 42, 48, 54],
        [60, 66, 72, 78, 84]],

       [[ 0,  7, 14, 21, 28],
        [35, 42, 49, 56, 63],
        [70, 77, 84, 91, 98]]]])
```

Let us unpack this one at a time and see what broadcasting is doing with each multiplication,

```
>>> X[0,0]*Y # 1st array element
array([[0, 0, 0, 0, 0],
       [0, 0, 0, 0, 0],
       [0, 0, 0, 0, 0]])
>>> X[0,1]*Y # 2nd array element
array([[ 0,  1,  2,  3,  4],
       [ 5,  6,  7,  8,  9],
       [10, 11, 12, 13, 14]])
>>> X[0,2]*Y # 3rd array element
array([[ 0,  2,  4,  6,  8],
       [10, 12, 14, 16, 18],
       [20, 22, 24, 26, 28]])
```

We can sum the items along any dimension with the `axis` keyword argument,

```
>>> (X[:,:,None,None]*Y).sum(axis=3) # sum along 4th dimension
array([[[  0,   0,   0],
        [ 10,  35,  60],
        [ 20,  70, 120],
        [ 30, 105, 180]],

       [[ 40, 140, 240],
        [ 50, 175, 300],
        [ 60, 210, 360],
        [ 70, 245, 420]]])
```

Abbreviating Loops with Broadcasting
Using broadcasting, compute the number of ways a quarter (i.e., 25 cents) can be split into pennies, nickels, and dimes:

```
>>> n=0 # start counter
>>> for n_d in range(0,3): # at most 2 dimes
...     for n_n in range(0,6): # at most 5 nickels
...         for n_p in range(0,26): # at most 25 pennies
...             value = n_d*10+n_n*5+n_p
...             if value == 25:
...                 print('dimes=%d, nickels=%d, pennies=%d'%(n_d,
                                                         n_n,n_p))
...                 n+=1
```

(continued)

```
...
dimes=0, nickels=0, pennies=25
dimes=0, nickels=1, pennies=20
dimes=0, nickels=2, pennies=15
dimes=0, nickels=3, pennies=10
dimes=0, nickels=4, pennies=5
dimes=0, nickels=5, pennies=0
dimes=1, nickels=0, pennies=15
dimes=1, nickels=1, pennies=10
dimes=1, nickels=2, pennies=5
dimes=1, nickels=3, pennies=0
dimes=2, nickels=0, pennies=5
dimes=2, nickels=1, pennies=0
>>> print('n = ',n)
n =  12
```

This can be done in *one* line with broadcasting:

```
>>> n_d = np.arange(3)
>>> n_n = np.arange(6)
>>> n_p = np.arange(26)
>>> # matches n above
>>> (n_p + 5*n_n[:,None] + 10*n_d[:,None,None]==25).sum()
12
```

This means the nested `for` loops above are equivalent to Numpy broadcasting so whenever you see this nested loop pattern, it may be an opportunity for broadcasting.

4.13 Masked Arrays

Numpy also allows for masking sections of Numpy arrays. This is very popular in image processing:

```
>>> x = np.array([2, 1, 3, np.nan, 5, 2, 3, np.nan])
>>> x
array([ 2.,  1.,  3., nan,  5.,  2.,  3., nan])
>>> np.mean(x)
nan
>>> m = np.ma.masked_array(x, np.isnan(x))
>>> m
masked_array(data=[2.0, 1.0, 3.0, --, 5.0, 2.0, 3.0, --],
             mask=[False, False, False,  True, False, False,
             False,  True],
       fill_value=1e+20)
>>> np.mean(m)
2.6666666666666665
>>> m.shape
(8,)
>>> x.shape
```

```
(8,)
>>> m.fill_value=9999
>>> m.filled()
array([2.000e+00, 1.000e+00, 3.000e+00, 9.999e+03, 5.000e+00,
2.000e+00,
       3.000e+00, 9.999e+03])
```

Making Numpy Arrays from Custom Objects
To make custom objects compatible with Numpy arrays, we have to define the
__array__ method:

```
>>> from numpy import arange
>>> class Foo():
...      def __init__(self): # note the double underscores
...           self.size = 10
...      def __array__(self): # produces numpy array
...           return arange(self.size)
...
>>> np.array(Foo())
array([0, 1, 2, 3, 4, 5, 6, 7, 8, 9])
```

4.14   Floating Point Numbers

There are precision limitations when representing floating point numbers on a
computer with finite memory. For example, the following shows these limitations
when adding two simple numbers,

```
>>> 0.1 + 0.2
0.30000000000000004
```

Why is the output not 0.3? The issue is the floating point representation of the two
numbers and the algorithm that adds them. To represent an integer in binary, we just
write it out in powers of 2. For example, $230 = (11100110)_2$. Python can do this
conversion using string formatting,

```
>>> '{0:b}'.format(230)
'11100110'
```

To add integers, we just add up the corresponding bits and fit them into the allowable
number of bits. Unless there is an overflow (the results cannot be represented with
that number of bits), then there is no problem. Representing floating point is trickier
because we have to represent these numbers as binary fractions. The IEEE 754
standard requires that floating point numbers to be represented as $\pm C \times 2^E$, where
C is the significand (*mantissa*) and E is the exponent.

To represent a regular decimal fraction as a binary fraction, we need to compute
the expansion of the fraction in the following form $a_1/2 + a_2/2^2 + a_3/2^3 \ldots$ In other
words, we need to find the a_i coefficients. We can do this using the same process we

would use for a decimal fraction: just keep dividing by the fractional powers of $1/2$ and keep track of the whole and fractional parts. Python's `divmod` function can do most of the work for this. For example, to represent `0.125` as a binary fraction,

```
>>> a = 0.125
>>> divmod(a*2,1)
(0.0, 0.25)
```

The first item in the tuple is the quotient and the other is the remainder. If the quotient was greater than 1, then the corresponding a_i term is one and is zero otherwise. For this example, we have $a_1 = 0$. To get the next term in the expansion, we just keep multiplying by 2, which moves us rightward along the expansion to a_{i+1} and so on. Then,

```
>>> a = 0.125
>>> q,a = divmod(a*2,1)
>>> (q,a)
(0.0, 0.25)
>>> q,a = divmod(a*2,1)
>>> (q,a)
(0.0, 0.5)
>>> q,a = divmod(a*2,1)
>>> (q,a)
(1.0, 0.0)
```

The algorithm stops when the remainder term is zero. Thus, we have that $0.125 = (0.001)_2$. The specification requires that the leading term in the expansion be one. Thus, we have $0.125 = (1.000) \times 2^{-3}$. This means the significand is 1 and the exponent is -3.

Now, let us get back to our main problem `0.1+0.2` by developing the representation `0.1` by coding up the individual steps above:

```
>>> a = 0.1
>>> bits = []
>>> while a>0:
...     q,a = divmod(a*2,1)
...     bits.append(q)
...
>>> ''.join(['%d'%i for i in bits])
'0001100110011001100110011001100110011001100110011001101'
```

Note that the representation has an infinitely repeating pattern. This means that we have $(1.\overline{1001})_2 \times 2^{-4}$. The IEEE standard does not have a way to represent infinitely repeating sequences. Nonetheless, we can compute this:

$$\sum_{n=1}^{\infty} \frac{1}{2^{4n-3}} + \frac{1}{2^{4n}} = \frac{3}{5}$$

Thus, $0.1 \approx 1.6 \times 2^{-4}$. Per the IEEE 754 standard, for `float` type, we have 24 bits for the significand and 23 bits for the fractional part. Because we cannot represent the infinitely repeating sequence, we have to round off at 23 bits,

1001100110011001101. Thus, whereas the significand's representation used to be 1.6, with this rounding, it is now

```
>>> b = '10011001100110011001101'
>>> 1+sum([int(i)/(2**n) for n,i in enumerate(b,1)])
1.600000023841858
```

Thus, we now have $0.1 \approx 1.600000023841858 \times 2^{-4} = 0.10000000149011612$. For the 0.2 expansion, we have the same repeating sequence with a different exponent, so that we have $0.2 \approx 1.600000023841858 \times 2^{-3} = 0.20000000298023224$. To add 0.1+0.2 in binary, we must adjust the exponents until they match the higher of the two. Thus,

```
  0.1100110011001100110
+1.1001100110011001101
-------------------------
10.0110011001100110011
```

Now, the sum has to be scaled back to fit into the significand's available bits so the result is 1.0011001100110011001010 with exponent -2. Computing this in the usual way as shown below gives the result:

```
>>> k='00110011001100110011010'
>>> ('%0.12f'%((1+sum([int(i)/(2**n)
...                          for n,i in enumerate(k,1)]))/2**2))
'0.300000011921'
```

which matches what we get with numpy

```
>>> import numpy as np
>>> '%0.12f'%(np.float32(0.1) + np.float32(0.2))
'0.300000011921'
```

The entire process proceeds the same for 64-bit floats. Python has fractions and decimal modules that allow more exact number representations. The decimal module is particularly important for certain financial computations.

Round-off Error   Let us consider the example of adding 100,000,000 and 10 in 32-bit floating point.

```
>>> '{0:b}'.format(100000000)
'101111101011110000100000000'
```

This means that $100,000,000 = (1.01111101011110000100000000)_2 \times 2^{26}$. Likewise, $10 = (1.010)_2 \times 2^3$. To add these, we have to make the exponents match as in the following:

```
  1.01111101011110000100000000
+0.00000000000000000000001010
-------------------------------
  1.01111101011110000100001010
```

Now, we have to round off because we only have 23 bits to the right of the decimal point and obtain 1.01111101011110000010000, thus losing the trailing 10 bits. This effectively makes the decimal $10 = (1010)_2$ we started out with become $8 = (1000)_2$. Thus, using Numpy again,

```
>>> format(np.float32(100000000) + np.float32(10),'10.3f')
'100000008.000'
```

The problem here is that the order of magnitude between the two numbers was so great that it resulted in loss in the significand's bits as the smaller number was right-shifted. When summing numbers like these, the Kahan summation algorithm (see `math.fsum()`) can effectively manage these round-off errors:

```
>>> import math
>>> math.fsum([np.float32(100000000),np.float32(10)])
100000010.0
```

Cancellation Error   Cancellation error (loss of significance) results when two nearly equal floating point numbers are subtracted. Let us consider subtracting `0.1111112` and `0.1111111`. As binary fractions, we have the following:

```
    1.11000111000111001000101 E-4
   -1.11000111000111000110111 E-4
   --------------------------
    0.00000000000000000011100
```

As a binary fraction, this is `1.11` with exponent `-23` or $(1.75)_{10} \times 2^{-23} \approx$ 0.00000010430812836. In Numpy, this loss of precision is shown in the following:

```
>>> format(np.float32(0.1111112)-np.float32(0.1111111),'1.17f')
'0.00000010430812836'
```

To sum up, when using floating point, you must check for approximate equality using something like Numpy `allclose` instead of the usual Python equality (i.e., `==`) sign. This enforces error bounds instead of strict equality. Whenever practicable, use fixed scaling to employ integer values instead of decimal fractions. Double precision 64-bit floating point numbers are much better than single precision and, while not eliminating these problems, effectively kick the can down the road for all but the strictest precision requirements. The Kahan algorithm is effective for summing floating point numbers across very large data without accruing round-off errors. To minimize cancellation errors, re-factor the calculation to avoid subtracting two nearly equal numbers.

Programming Tip: `decimal` Module
Python has a built-in `decimal` module that uses a different method to manage floating point numbers, especially for financial calculations. The downside is that `decimal` is much slower than the IEEE standard for floating point described here.

4.15 Advanced Numpy `dtypes`

Numpy `dtypes` can also help read structured sections of binary data files. For example, a WAV file has a 44-byte header format:

```
Item                    Description
-------------------     ------------------------------------------
chunk_id                "RIFF"
chunk_size              4-byte unsigned little-endian integer
format                  "WAVE"
fmt_id                  "fmt"
fmt_size                4-byte unsigned little-endian integer
audio_fmt               2-byte unsigned little-endian integer
num_channels            2-byte unsigned little-endian integer
sample_rate             4-byte unsigned little-endian integer
byte_rate               4-byte unsigned little-endian integer
block_align             2-byte unsigned little-endian integer
bits_per_sample      2-byte unsigned little-endian integer
data_id                 "data"
data_size               4-byte unsigned little-endian integer
```

You can open this test sample file off the web using the following code to get the first four bytes of the `chunk_id`:

```
>>> from urllib.request import urlopen
>>> fp=urlopen('https://www.kozco.com/tech/piano2.wav')
>>> fp.read(4)
b'RIFF'
```

Reading the next four little-endian bytes gives the `chunk_size`:

```
>>> fp.read(4)
b'\x04z\x12\x00'
```

Note that there is an issue of the endianness of the returned bytes. We can get next four bytes as follows:

```
>>> fp.read(4)
b'WAVE'
```

Continuing like this, we could get the rest of the header, but it is better to use the following custom `dtype`:

```
>>> header_dtype = np.dtype([
...     ('chunkID','S4'),
...     ('chunkSize','<u4'),
...     ('format','S4'),
...     ('subchunk1ID','S4'),
...     ('subchunk1Size','<u4'),
...     ('audioFormat','<u2'),
...     ('numChannels','<u2'),
...     ('sampleRate','<u4'),
...     ('byteRate','<u4'),
...     ('blockAlign','<u2'),
...     ('bitsPerSample','<u2'),
...     ('subchunk2ID','S4'),
...     ('subchunk2Size','u4'),
...     ])
```

Then, we start all over and do the following but with `urlretrieve`, which writes the temporary file that Numpy needs:

```
>>> from urllib.request import urlretrieve
>>> path, _ = urlretrieve('https://www.kozco.com/tech/piano2.wav')
>>> h=np.fromfile(path,dtype=header_dtype,count=1)
>>> print(h)
[(b'RIFF', 1210884, b'WAVE', b'fmt ', 16, 1, 2, 48000, 192000, 4,
16, b'data', 1210848)]
```

Note that the output neatly encapsulates the individual bytes with their corresponding endian issues. Now that we have the header that tells us how many bytes are in the data, and we can read the rest of the data and process it correctly using the other header fields.

> **Formatting Numpy Arrays**
> Sometimes Numpy arrays can get very dense to view. The `np.set_printoptions` function can help reduce the visual clutter by providing custom formatting options for the different Numpy data types. Keep in mind that is a formatting issue and none of the underlying data is changed. For example, to change the representation of floating point Numpy arrays to `%3.2f`, we can do the following:
>
> ```
> np.set_printoptions(formatter={'float':lambda i:'%3.2f'%i}
> ```

References

1. T.E. Oliphant, *A Guide to NumPy* (Trelgol Publishing, Austin, 2006)

Chapter 5
Pandas

Pandas is a powerful module that is optimized on top of Numpy and provides a set of data structures particularly suited to time-series and spreadsheet-style data analysis (think of pivot tables in Excel). If you are familiar with the R statistical package, then you can think of Pandas as providing a Numpy-powered DataFrame for Python. Pandas provides a DataFrame object (among others) built on a Numpy platform to ease data manipulation (especially for time-series) for statistical processing. Pandas is particularly popular in quantitative finance. Key features of Pandas include fast data manipulation and alignment, tools for exchanging data between different formats and between SQL databases, handling missing data, and cleaning up messy data.

5.1 Using Series

The easiest way to think about Pandas Series objects is as a container for two Numpy arrays, one for the index and the other for the data. Recall that Numpy arrays already have integer-indexing just like regular Python lists.

```
>>> import pandas as pd
>>> x = pd.Series([1,2,30,0,15,6])
>>> x
0     1
1     2
2    30
3     0
4    15
5     6
dtype: int64
```

This object can be indexed in plain Numpy style,

```
>>> x[1:3] # Numpy slicing
1     2
```

© The Editor(s) (if applicable) and The Author(s), under exclusive license to Springer Nature Switzerland AG 2021
J. Unpingco, *Python Programming for Data Analysis*,
https://doi.org/10.1007/978-3-030-68952-0_5

```
2      30
dtype: int64
```

We can also get the Numpy arrays directly,

```
>>> x.values # values
array([ 1,   2, 30,   0, 15,   6])
>>> x.values[1:3]
array([ 2, 30])
>>> x.index   # index is Numpy array-like
RangeIndex(start=0, stop=6, step=1)
```

Unlike Numpy arrays, you can have mixed types,

```
>>> s = pd.Series([1,2,'anything','more stuff'])
>>> s
0               1
1               2
2       anything
3     more stuff
dtype: object
>>> s.index # Series index
RangeIndex(start=0, stop=4, step=1)
>>> s[0] # The usual Numpy slicing rules apply
1
>>> s[:-1]
0               1
1               2
2       anything
dtype: object
>>> s.dtype   # object data type
dtype('O')
```

Beware that mixed types in a single column can lead to downstream inefficiencies and other problems. The index in the `pd.Series` generalizes beyond integer-indexing. For example,

```
>>> s = pd.Series([1,2,3],index=['a','b','cat'])
>>> s['a']
1
>>> s['cat']
3
```

Because of its legacy as a financial data (i.e., stock prices) processing tool, Pandas is *really* good at managing time-series

```
>>> dates = pd.date_range('20210101',periods=12)
>>> s = pd.Series(range(12),index=dates) # explicitly assign index
>>> s # default is calendar-daily
2021-01-01     0
2021-01-02     1
2021-01-03     2
2021-01-04     3
2021-01-05     4
2021-01-06     5
2021-01-07     6
```

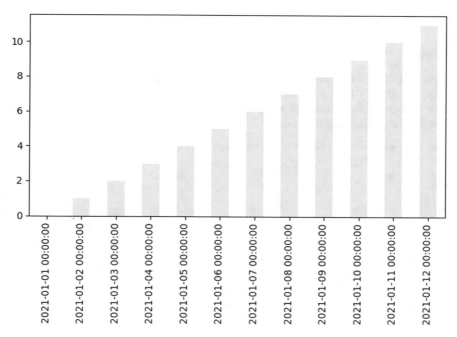

Fig. 5.1 Quick plot of `Series` object

```
2021-01-08      7
2021-01-09      8
2021-01-10      9
2021-01-11     10
2021-01-12     11
Freq: D, dtype: int64
```

You can do some basic descriptive statistics on the data (not the index!) right away

```
>>> s.mean()
5.5
>>> s.std()
3.605551275463989
```

You can also plot (see Fig. 5.1) the `Series` using its methods:

```
>>> s.plot(kind='bar',alpha=0.3) # can add extra matplotlib
↪    keywords
```

Data can be summarized by the index. For example, to count the individual days of the week for which we had data:

```
>>> s.groupby(by=lambda i:i.dayofweek).count()
0      2
1      2
2      1
3      1
```

```
4     2
5     2
6     2
dtype: int64
```

Note that the convention 0 is Monday, 1 is Tuesday, and so forth. Thus, there was only one Sunday (day 6) in the dataset. The groupby method partitions the data into disjoint groups based on the predicate given via the by keyword argument. Consider the following Series,

```
>>> x = pd.Series(range(5),index=[1,2,11,9,10])
>>> x
1     0
2     1
11    2
9     3
10    4
dtype: int64
```

Let us group it in the following according to whether the elements in the values are even or odd using the modulus (%) operator,

```
>>> grp = x.groupby(lambda i:i%2) # odd or even
>>> grp.get_group(0) # even group
2     1
10    4
dtype: int64
>>> grp.get_group(1) # odd group
1     0
11    2
9     3
dtype: int64
```

The first line groups the elements of the Series object by whether or not the index is even or odd. The lambda function returns 0 or 1 depending on whether or not the corresponding index is even or odd, respectively. The next line shows the 0 (i.e., even) group and then the one after shows the 1 (odd) group. Now that we have separate groups, and we can perform a wide variety of summarizations on each to reduce it to a single value. For example, in the following, we get the maximum value of each group:

```
>>> grp.max() # max in each group
0     4
1     3
dtype: int64
```

Note that the operation above returns another Series object with an index corresponding to the [0,1] elements. There will be as many groups as there are unique outputs of the by function.

5.2 Using DataFrame

While the `Series` object can be thought of as encapsulating two Numpy arrays (index and values), the Pandas `DataFrame` is an encapsulation of group of `Series` objects that share a *single* index. We can create a `DataFrame` with dictionaries as in the following:

```
>>> df = pd.DataFrame({'col1': [1,3,11,2], 'col2': [9,23,0,2]})
>>> df
   col1  col2
0     1     9
1     3    23
2    11     0
3     2     2
```

Note that the keys in the input dictionary are now the column headings (labels) of the `DataFrame`, with each corresponding column matching the list of corresponding values from the dictionary. Like the `Series` object, the `DataFrame` also has an index, which is the `[0,1,2,3]` column on the far left. We can extract elements from each column using the `iloc`, which ignore the given index and return to traditional Numpy slicing,

```
>>> df.iloc[:2,:2] # get section
   col1  col2
0     1     9
1     3    23
```

or by directly slicing or by using the *dot* notation as shown below:

```
>>> df['col1'] # indexing
0     1
1     3
2    11
3     2
Name: col1, dtype: int64
>>> df.col1 # use dot notation
0     1
1     3
2    11
3     2
Name: col1, dtype: int64
```

> **Programming Tip: Spaces in Column Names**
> As long as the names of the columns in the `DataFrame` do *not* contain spaces or other `eval`-able syntax like hyphens, you can use the dot notation attribute-style access to the column values. You can report the column names using `df.columns`.

Subsequent operations on the `DataFrame` preserve its column-wise structure as in the following:

```
>>> df.sum()
col1     17
col2     34
dtype: int64
```

where each column was totaled. Grouping and aggregating with the DataFrame is even more powerful than with Series. Let us construct the following DataFrame,

```
>>> df = pd.DataFrame({'col1': [1,1,0,0], 'col2': [1,2,3,4]})
>>> df
   col1   col2
0     1      1
1     1      2
2     0      3
3     0      4
```

In the above DataFrame, note that the col1 column has only two *distinct* entries. We can group the data using this column as in the following:

```
>>> grp=df.groupby('col1')
>>> grp.get_group(0)
   col1   col2
2     0      3
3     0      4
>>> grp.get_group(1)
   col1   col2
0     1      1
1     1      2
```

Note that each group corresponds to entries for which col1 was either of its two values. Now that we have grouped on col1, as with the Series object, we can also functionally summarize each of the groups as in the following:

```
>>> grp.sum()
       col2
col1
0         7
1         3
```

where the sum is applied across each of the dataframes present in each group. Note that the index of the output above is each of the values in the original col1.

The DataFrame can compute new columns based on the existing columns using the eval method as shown below:

```
>>> df['sum_col']=df.eval('col1+col2')
>>> df
   col1   col2   sum_col
0     1      1         2
1     1      2         3
2     0      3         3
3     0      4         4
```

Note that you can assign the output to a new column to the DataFrame as shown. We can group by multiple columns as shown below:

```
>>> grp = df.groupby(['sum_col','col1'])
```

Doing the sum operation on each group gives the following:

```
>>> res = grp.sum()
>>> res
                  col2
sum_col col1
2       1         1
3       0         3
        1         2
4       0         4
```

This output is much more complicated than anything we have seen so far, so let us carefully walk through it. Below the headers, the first row 2 1 1 indicates that for sum_col=2 and for all values of col1 (namely, just the value 1), the value of col2 is 1. For the next row, the same pattern applies except that for sum_col=3, there are now two values for col1, namely 0 and 1, which each have their corresponding two values for the sum operation in col2. This layered display is one way to look at the result. Note that the layers above are not uniform. Alternatively, we can unstack this result to obtain the following tabular view of the previous result:

```
>>> res.unstack()
         col2
col1        0    1
sum_col
2         NaN  1.0
3         3.0  2.0
4         4.0  NaN
```

The NaN values indicate positions in the table where there is no entry. For example, for the pair (sum_col=2, col2=0), there is no corresponding value in the DataFrame, as you may verify by looking at the penultimate code block. There is also no entry corresponding to the (sum_col=4, col2=1) pair. Thus, this shows that the original presentation in the penultimate code block is the same as this one, just without the above-mentioned missing entries indicated by NaN.

Let us continue with indexing dataframes.

```
>>> import numpy as np
>>> data=np.arange(len(dates)*4).reshape(-1,4)
>>> df = pd.DataFrame(data,index=dates,
...                              columns=['A','B','C','D'])
>>> df
             A    B    C    D
2021-01-01   0    1    2    3
2021-01-02   4    5    6    7
2021-01-03   8    9   10   11
2021-01-04  12   13   14   15
2021-01-05  16   17   18   19
2021-01-06  20   21   22   23
2021-01-07  24   25   26   27
2021-01-08  28   29   30   31
2021-01-09  32   33   34   35
2021-01-10  36   37   38   39
```

```
2021-01-11   40   41   42   43
2021-01-12   44   45   46   47
```

Now, you can access each of the columns by name, as in the following:

```
>>> df['A']
2021-01-01      0
2021-01-02      4
2021-01-03      8
2021-01-04     12
2021-01-05     16
2021-01-06     20
2021-01-07     24
2021-01-08     28
2021-01-09     32
2021-01-10     36
2021-01-11     40
2021-01-12     44
Freq: D, Name: A, dtype: int64
```

Or, using a quicker attribute-style notation

```
>>> df.A
2021-01-01      0
2021-01-02      4
2021-01-03      8
2021-01-04     12
2021-01-05     16
2021-01-06     20
2021-01-07     24
2021-01-08     28
2021-01-09     32
2021-01-10     36
2021-01-11     40
2021-01-12     44
Freq: D, Name: A, dtype: int64
```

Now, we can do some basic computing and indexing.

```
>>> df.loc[:dates[3]] # unlike the Python convention, this
↪   includes endpoints!
             A    B    C    D
2021-01-01   0    1    2    3
2021-01-02   4    5    6    7
2021-01-03   8    9   10   11
2021-01-04  12   13   14   15
>>> df.loc[:,'A':'C'] # all rows by slice of column labels
             A    B    C
2021-01-01   0    1    2
2021-01-02   4    5    6
2021-01-03   8    9   10
2021-01-04  12   13   14
2021-01-05  16   17   18
2021-01-06  20   21   22
2021-01-07  24   25   26
2021-01-08  28   29   30
2021-01-09  32   33   34
```

```
2021-01-10   36   37   38
2021-01-11   40   41   42
2021-01-12   44   45   46
```

5.3 Reindexing

A `DataFrame` or `Series` has an index that can align data using reindexing,

```
>>> x = pd.Series(range(3),index=['a','b','c'])
>>> x
a    0
b    1
c    2
dtype: int64
>>> x.reindex(['c','b','a','z'])
c    2.0
b    1.0
a    0.0
z    NaN
dtype: float64
```

Note how the newly created Series object has a new index and fills in missing items with NaN. You can fill in by other values by using the `fill_value` keyword argument in reindex. It is also possible to do back-filling and forward-filling (`ffill`) of values when working with ordered data as in the following:

```
>>> x = pd.Series(['a','b','c'],index=[0,5,10])
>>> x
0     a
5     b
10    c
dtype: object
>>> x.reindex(range(11),method='ffill')
0     a
1     a
2     a
3     a
4     a
5     b
6     b
7     b
8     b
9     b
10    c
dtype: object
```

More complicated interpolation methods are possible, but not directly using reindex. Reindexing also applies to dataframes, but in either or both of the two dimensions.

```
>>> df = pd.DataFrame(index=['a','b','c'],
...                   columns=['A','B','C','D'],
...                   data = np.arange(3*4).reshape(3,4))
```

```
>>> df
   A   B    C    D
a  0   1    2    3
b  4   5    6    7
c  8   9   10   11
```

Now, we can reindex this by the index as in the following:

```
>>> df.reindex(['c','b','a','z'])
     A     B      C      D
c  8.0   9.0   10.0   11.0
b  4.0   5.0    6.0    7.0
a  0.0   1.0    2.0    3.0
z  NaN   NaN    NaN    NaN
```

Note how the missing z element has been filled in as with the prior Series object. The same behavior applies to reindexing the columns as in the following:

```
>>> df.reindex(columns=['D','A','C','Z','B'])
    D   A    C    Z   B
a   3   0    2  NaN   1
b   7   4    6  NaN   5
c  11   8   10  NaN   9
```

Again, we have the same filling behavior, now just applied to the columns. The same back-filling and forward-filling works with ordered columns/indices as before.

5.4   Deleting Items

Returning to our previous Series object,

```
>>> x = pd.Series(range(3),index=['a','b','c'])
```

To get rid of the data at the ′a′ index, we can use the drop method that will return a new Series object with the specified data removed,

```
>>> x.drop('a')
b    1
c    2
dtype: int64
```

Bear in mind this is a *new* Series object unless we use the inplace keyword argument or explicitly using del,

```
>>> del x['a']
```

The same pattern applies to DataFrames.

```
>>> df = pd.DataFrame(index=['a','b','c'],
...                   columns=['A','B','C','D'],
...                   data = np.arange(3*4).reshape(3,4))
>>> df.drop('a')
   A   B    C    D
b  4   5    6    7
c  8   9   10   11
```

Or, along the column dimension,

```
>>> df.drop('A',axis=1)
   B   C   D
a  1   2   3
b  5   6   7
c  9  10  11
```

Again, the same comments regarding using `del` and `inplace` also apply for DataFrames.

5.5 Advanced Indexing

Pandas provides very powerful and fast slicing,

```
>>> x = pd.Series(range(4),index=['a','b','c','d'])
>>> x['a':'c']
a    0
b    1
c    2
dtype: int64
```

Note that unlike regular Python indexing, *both* endpoints are included here when slicing with labels. This can also be used to assign values as in the following:

```
>>> x['a':'c']=999
>>> x
a    999
b    999
c    999
d      3
dtype: int64
```

Analogous behavior applies to DataFrames.

```
>>> df = pd.DataFrame(index=['a','b','c'],
...                   columns=['A','B','C','D'],
...                   data = np.arange(3*4).reshape(3,4))
>>> df['a':'b']
   A  B  C  D
a  0  1  2  3
b  4  5  6  7
```

You can pick out individual columns without the colon (`:`).

```
>>> df[['A','C']]
   A   C
a  0   2
b  4   6
c  8  10
```

Mixing label-based slices with Numpy-like colon slicing is possible using `loc`,

```
>>> df.loc['a':'b',['A','C']]
   A  C
a  0  2
b  4  6
```

The idea is that the first argument to `loc` indexes the rows and the second indexes the columns. Heuristics allow for Numpy-like indexing without worrying about the labels. You can go back to plain Numpy indexing with `iloc`.

```
>>> df.iloc[0,-2:]
C    2
D    3
Name: a, dtype: int64
```

5.6   Broadcasting and Data Alignment

The main thing to keep in mind when operating on one or more `Series` or `DataFrame` objects is that the `index` always aligns the computation.

```
>>> x = pd.Series(range(4),index=['a','b','c','d'])
>>> y = pd.Series(range(3),index=['a','b','c'])
```

Note that `y` is missing one of the indices in `x`, so when we add them,

```
>>> x+y
a    0.0
b    2.0
c    4.0
d    NaN
dtype: float64
```

Note that because `y` was missing one of the indices, it was filled in with a `NaN`. This behavior also applies to dataframes,

```
>>> df = pd.DataFrame(index=['a','b','c'],
...                   columns=['A','B','C','D'],
...                   data = np.arange(3*4).reshape(3,4))
>>> ef = pd.DataFrame(index=list('abcd'),
...                   columns=list('ABCDE'),
...                   data = np.arange(4*5).reshape(4,5))
>>> ef
    A   B   C   D   E
a   0   1   2   3   4
b   5   6   7   8   9
c  10  11  12  13  14
d  15  16  17  18  19
>>> df
   A  B   C   D
a  0  1   2   3
b  4  5   6   7
c  8  9  10  11
>>> df+ef
      A      B      C      D   E
```

```
a    0.0    2.0    4.0    6.0 NaN
b    9.0   11.0   13.0   15.0 NaN
c   18.0   20.0   22.0   24.0 NaN
d    NaN    NaN    NaN    NaN NaN
```

Note that the non-overlapping elements are filled in with NaN. For simple operations, you can specify the fill value using the named operation. For example, in the last case,

```
>>> df.add(ef,fill_value=0)
        A      B      C      D      E
a    0.0    2.0    4.0    6.0    4.0
b    9.0   11.0   13.0   15.0    9.0
c   18.0   20.0   22.0   24.0   14.0
d   15.0   16.0   17.0   18.0   19.0

>>> s = df.loc['a'] # take first row
>>> s
A    0
B    1
C    2
D    3
Name: a, dtype: int64
```

When we add this Series object with the full DataFrame, we obtain the following:

```
>>> s + df
   A   B   C   D
a  0   2   4   6
b  4   6   8  10
c  8  10  12  14
```

Compare this to the original DataFrame,

```
>>> df
   A  B   C   D
a  0  1   2   3
b  4  5   6   7
c  8  9  10  11
```

This shows that the Series object was broadcast down the rows, aligning with the columns in the DataFrame. Here is an example of a different Series object that is missing some of the columns in the DataFrame,

```
>>> s = pd.Series([1,2,3],index=['A','D','E'])
>>> s+df
     A    B   C     D   E
a  1.0  NaN NaN   5.0 NaN
b  5.0  NaN NaN   9.0 NaN
c  9.0  NaN NaN  13.0 NaN
```

Note that the broadcasting still happens down the rows, aligning with the columns, but fills in the missing entries with NaN.

Here is a quick Python test that uses regular expressions to test for relatively small prime numbers,

```
>>> import re
>>> pattern = r'^1?$|^(11+?)\1+$'
>>> def isprime(n):
...     return (re.match(pattern, '1'*n) is None) #*
...
```

Now, we can find which column label has the most prime numbers in it

```
>>> df.applymap(isprime)
       A      B      C      D
a  False  False   True   True
b  False   True  False   True
c  False  False  False   True
```

The booleans are automatically cast in the sum below:

```
>>> df.applymap(isprime).sum()
A    0
B    1
C    1
D    3
dtype: int64
```

This just scratches the surface of the kinds of fluid data analysis that are almost automatic using Pandas.

> **Programming Tip: Pandas Performance**
> Pandas `groupby`, `apply`, and `applymap` are flexible and powerful, but Pandas has to evaluate them in the Python interpreter and *not* in the optimized Pandas code, which results in significant slowdown. Thus, it is always best to use the functions that are built into Pandas itself instead defining pure Python functions to feed into these methods.

5.7 Categorical and Merging

Pandas supports a few relational algebraic operations like table joins.

```
>>> df = pd.DataFrame(index=['a','b','c'],
...                   columns=['A','B','C'],
...                   data = np.arange(3*3).reshape(3,3))
>>> df
   A  B  C
a  0  1  2
b  3  4  5
c  6  7  8
>>> ef = pd.DataFrame(index=['a','b','c'],
...                   columns=['A','Y','Z'],
...                   data = np.arange(3*3).reshape(3,3))
>>> ef
```

```
     A   Y   Z
a    0   1   2
b    3   4   5
c    6   7   8
```

The table join is implemented in the `merge` function.

```
>>> pd.merge(df,ef,on='A')
     A   B   C   Y   Z
0    0   1   2   1   2
1    3   4   5   4   5
2    6   7   8   7   8
```

The `on` keyword argument says to merge the two DataFrames where they have matching corresponding entries in the `A` column. Note that the index was not preserved in the merge. To make things more interesting, let us make the `ef` DataFrame different by dropping one of the rows,

```
>>> ef.drop('b',inplace=True)
>>> ef
     A   Y   Z
a    0   1   2
c    6   7   8
```

Now, let us try the `merge` again.

```
>>> pd.merge(df,ef,on='A')
     A   B   C   Y   Z
0    0   1   2   1   2
1    6   7   8   7   8
```

Note that only the elements of `A` that match both DataFrames (i.e., are in the intersection of both) are preserved. We can alter this by using the `how` keyword argument.

```
>>> pd.merge(df,ef,on='A',how='left')
     A   B   C    Y     Z
0    0   1   2   1.0   2.0
1    3   4   5   NaN   NaN
2    6   7   8   7.0   8.0
```

The `how=left` keyword argument tells the join to keep all the keys on the left DataFrame (`df` in this case) and fill in with `NaN` wherever missing in the right DataFrame (`ef`). If `ef` has elements along `A` that are absent in `df`, then these would disappear,

```
>>> ef = pd.DataFrame(index=['a','d','c'],
...                   columns=['A','Y','Z'],
...                   data = 10*np.arange(3*3).reshape(3,3))
>>> ef
     A    Y    Z
a    0   10   20
d   30   40   50
c   60   70   80
```

```
>>> pd.merge(df,ef,on='A',how='left')
   A  B  C    Y      Z
0  0  1  2  10.0   20.0
1  3  4  5  NaN    NaN
2  6  7  8  NaN    NaN
```

Likewise, we can do how=right to use the right DataFrame keys,

```
>>> pd.merge(df,ef,on='A',how='right')
    A    B    C   Y   Z
0   0  1.0  2.0  10  20
1  30  NaN  NaN  40  50
2  60  NaN  NaN  70  80
```

We can use how=outer to get the union of the keys,

```
>>> pd.merge(df,ef,on='A',how='outer')
    A    B    C    Y      Z
0   0  1.0  2.0  10.0   20.0
1   3  4.0  5.0  NaN    NaN
2   6  7.0  8.0  NaN    NaN
3  30  NaN  NaN  40.0   50.0
4  60  NaN  NaN  70.0   80.0
```

Another common task is to split continuous data into discrete bins.

```
>>> a = np.arange(10)
>>> a
array([0, 1, 2, 3, 4, 5, 6, 7, 8, 9])
>>> bins = [0,5,10]
>>> cats = pd.cut(a,bins)
>>> cats
[NaN, (0.0, 5.0], (0.0, 5.0], (0.0, 5.0], (0.0, 5.0], (0.0, 5.0],
(5.0, 10.0], (5.0, 10.0], (5.0, 10.0], (5.0, 10.0]]
Categories (2, interval[int64]): [(0, 5] < (5, 10]]
```

The pd.cut function takes the data in the array and puts them into the categorical
variable cats.

```
>>> cats.categories
IntervalIndex([(0, 5], (5, 10]],
              closed='right',
              dtype='interval[int64]')
```

The half-open intervals indicate the bounds of each category. You can change the
parity of the intervals by passing the right=False keyword argument.

```
>>> cats.codes
array([-1,  0,  0,  0,  0,  0,  1,  1,  1,  1], dtype=int8)
```

The -1 above means that the 0 is not included in either of the two categories
because the interval is open on the left. You can count the number of elements in
each category as shown next,

```
>>> pd.value_counts(cats)
(0, 5]      5
(5, 10]     4
dtype: int64
```

Descriptive names for each category can be passed using the `labels` keyword argument.

```
>>> cats = pd.cut(a,bins,labels=['one','two'])
>>> cats
[NaN, 'one', 'one', 'one', 'one', 'one', 'two', 'two', 'two',
↪  'two']
Categories (2, object): ['one' < 'two']
```

Note that if you pass an integer argument for `bins`, it will automatically split into equal-sized categories. The `qcut` function is very similar except that it splits on quartiles.

```
>>> a = np.random.rand(100) # uniform random variables
>>> cats = pd.qcut(a,4,labels=['q1','q2','q3','q4'])
>>> pd.value_counts(cats)
q4    25
q3    25
q2    25
q1    25
dtype: int64
```

5.8 Memory Usage and `dtypes`

You will find yourself processing a *lot* of data with Pandas. Here are some tips to do that efficiently. First, we need the Penguins dataset from Seaborn,

```
>>> import seaborn as sns
>>> df = sns.load_dataset('penguins')
>>> df.head()
  species     island bill_length_mm bill_depth_mm flipper_length_mm body_mass_g     sex
0  Adelie  Torgersen          39.1          18.7             181.0       3750.0    Male
1  Adelie  Torgersen          39.5          17.4             186.0       3800.0  Female
2  Adelie  Torgersen          40.3          18.0             195.0       3250.0  Female
3  Adelie  Torgersen           NaN           NaN               NaN          NaN     NaN
4  Adelie  Torgersen          36.7          19.3             193.0       3450.0  Female
```

This is not a particularly big dataset, but it will suffice. Let us examine the `dtypes` of the `DataFrame`,

```
>>> df.dtypes
species              object
island               object
bill_length_mm      float64
bill_depth_mm       float64
flipper_length_mm   float64
body_mass_g         float64
sex                  object
dtype: object
```

Notice that some of these are marked `object`. This is usually means inefficiency because this generalized dtype may consume an inordinate amount of memory. Pandas comes with a simple way to assess the memory consumption of your `DataFrame`,

```
>>> df.memory_usage(deep=True)
Index                      128
species                  21876
island                   21704
bill_length_mm            2752
bill_depth_mm             2752
flipper_length_mm         2752
body_mass_g               2752
sex                      20995
dtype: int64
```

Now, we have an idea of our memory consumption for this `DataFrame`, and we can improve it by changing the dtypes. The categorical type we discussed previously can be specified as a new dtype for the `sex` column,

```
>>> ef = df.astype({'sex':'category'})
>>> ef.memory_usage(deep=True)
Index                      128
species                  21876
island                   21704
bill_length_mm            2752
bill_depth_mm             2752
flipper_length_mm         2752
body_mass_g               2752
sex                        548
dtype: int64
```

This results in almost a 40 times reduction in memory for this, which may be significant if the `DataFrame` had thousands of rows, for example. This works because there are many more rows than distinct values in the `sex` column. Let us continue using `category` as the dtype for the `species` and `island` columns.

```
>>> ef = df.astype({'sex':'category',
...                  'species':'category',
...                  'island':'category'})
>>> ef.memory_usage(deep=True)
Index                      128
species                    616
island                     615
bill_length_mm            2752
bill_depth_mm             2752
flipper_length_mm         2752
body_mass_g               2752
sex                        548
dtype: int64
```

To compare, we can put these side-by-side into a new `DataFrame`,

```
>>> (pd.DataFrame({'df':df.memory_usage(deep=True),
...                'ef':ef.memory_usage(deep=True)})
...     .assign(ratio= lambda i:i.ef/i.df))
                       df     ef      ratio
Index                 128    128   1.000000
species             21876    616   0.028159
island              21704    615   0.028336
bill_length_mm       2752   2752   1.000000
bill_depth_mm        2752   2752   1.000000
flipper_length_mm    2752   2752   1.000000
body_mass_g          2752   2752   1.000000
sex                 20995    548   0.026101
```

This shows a much smaller memory footprint for the columns that we changed to categorical dtype. We can also change the numerical types from the default `float64`, if we do not need that level of precision. For example, the `flipper_length_mm` column is measured in millimeters and there is no fractional part to any of the measurements. Thus, we can change that column as the following dtype and save four times the memory,

```
>>> ef = df.astype({'sex':'category',
...                 'species':'category',
...                 'island':'category',
...                 'flipper_length_mm': np.float16})
>>> ef.memory_usage(deep=True)
Index                 128
species               616
island                615
bill_length_mm       2752
bill_depth_mm        2752
flipper_length_mm     688
body_mass_g          2752
sex                   548
dtype: int64
```

Here is the summary again,

```
>>> (pd.DataFrame({'df':df.memory_usage(deep=True),
...                'ef':ef.memory_usage(deep=True)})
...     .assign(ratio= lambda i:i.ef/i.df))
                       df     ef      ratio
Index                 128    128   1.000000
species             21876    616   0.028159
island              21704    615   0.028336
bill_length_mm       2752   2752   1.000000
bill_depth_mm        2752   2752   1.000000
flipper_length_mm    2752    688   0.250000
body_mass_g          2752   2752   1.000000
sex                 20995    548   0.026101
```

Thus, by changing the default `object` dtype to other smaller dtypes can result in significant savings and potentially speed up downstream processing for dataframes. This is particularly true when scraping data from the web directly into dataframes using `pd.read_html`, which, even though the data looks numerical on the webpage, will typically result in the unnecessarily heavy `object` dtype.

5.9   Common Operations

A common problem is how to split a string column into components. For example,

```
>>> df = pd.DataFrame(dict(name=['Jon Doe','Jane Smith']))
>>> df.name.str.split(' ',expand=True)
      0      1
0   Jon    Doe
1  Jane  Smith
```

The key step is the `expand` keyword argument that converts the result into a DataFrame. The result can be assigned into the same DataFrame using the following,

```
>>> df[['first','last']]=df.name.str.split(' ',expand=True)
>>> df
         name first    last
0     Jon Doe   Jon     Doe
1  Jane Smith  Jane   Smith
```

Note that failing to use the `expand` keyword argument results in an output list instead of a DataFrame

```
>>> df.name.str.split(' ')
0        [Jon, Doe]
1      [Jane, Smith]
Name: name, dtype: object
```

This can be fixed by using `apply` on the output to convert from a `list` into a `Series` object as shown,[1]

```
>>> df.name.str.split(' ').apply(pd.Series)
      0      1
0   Jon    Doe
1  Jane  Smith
```

The `apply` method is one of the most powerful and general `DataFrame` methods. It operates on the individual columns (i.e., `pd.Series`) objects of the `DataFrame`. Unlike the `applymap` method that leaves the shape of the `DataFrame` unchanged, the `apply` method can return objects of a different shape. For example, doing something like `df.apply(lambda i:i[i>3])` on a `DataFrame` with numeric rows will return a small truncated NaN-filled `DataFrame`. Further, the `df.apply(raw=True)` keyword argument speeds

---

[1]There are many other Python string methods in the `.str` submodule such as `rstrip`, `upper`, and `title`.

up the method by operating directly on the underlying Numpy array in the `DataFrame` columns. This means that the `'apply'` method processes the Numpy arrays directly instead of the usual `'pd.Series'` objects.

The `transform` method is closely related to `apply` but must produce an output `DataFrame` of the same dimensions. For example,

```
>>> df = pd.DataFrame({'A': [1,1,2,2], 'B': range(4)})
>>> df
   A  B
0  1  0
1  1  1
2  2  2
3  2  3
```

We can group by `'A'` and reduce using the usual aggregation,

```
>>> df.groupby('A').sum()
   B
A
1  1
2  5
```

But by using `.transform()` we can broadcast the results to their respective places in the original `DataFrame`.

```
>>> df.groupby('A').transform('sum')
   B
0  1
1  1
2  5
3  5
```

The `describe` DataFrame method is useful for summarizing a given DataFrame, as shown below,

```
>>> df = pd.DataFrame(index=['a','b','c'], columns=['A','B','C'],
↪    data = np.arange(3*3).reshape(3,3))
>>> df.describe()
         A    B    C
count  3.0  3.0  3.0
mean   3.0  4.0  5.0
std    3.0  3.0  3.0
min    0.0  1.0  2.0
25%    1.5  2.5  3.5
50%    3.0  4.0  5.0
75%    4.5  5.5  6.5
max    6.0  7.0  8.0
```

It is often useful to get rid of accidental duplicated entries.

```
>>> df = pd.DataFrame({'A': [1,1,2,2,2,3], 'B': range(6)})
>>> df.drop_duplicates('A')
   A  B
0  1  0
2  2  2
5  3  5
```

The `keep` keyword argument decides which of the duplicated entries to retain.

5.10   Displaying DataFrames

Pandas has the `set_option` method to change the visual display of DataFrames while *not* altering the corresponding data elements.

```
>>> pd.set_option('display.float_format','{:.2f}'.format)
```

Note that the argument is a `callable` that produces the formatted string. These custom settings can be undone with `reset_option`, as in

```
pd.reset_option('display.float_format')
```

The `chop` option is handy for trimming excessive display precision,

```
>>> pd.set_option('display.chop',1e-5)
```

Within the Jupyter Notebook, formatting can utilize HTML elements with the `style.format` DataFrame method. For example, as shown in Fig. 5.2,

```
>>> from pandas_datareader import data
>>> df=data.DataReader("F", 'yahoo', '20200101',
↪    '20200110').reset_index()
>>> (df.style.format(dict(Date='{:%m/%d/%Y}'))
...   .hide_index()
...   .highlight_min('Close',color='red')
...   .highlight_max('Close',color='lightgreen')
... )
<pandas.io.formats.style.Styler object at 0x7f9376a22460>
```

formats the resulting table with the minimum closing price highlighted in red and the maximum closing price highlighted in green. Providing these kinds of quick visual cues is critically important for picking out key data elements. Note that the parenthesis above is for using the newlines to separate the `dot` methods, which is a style inherited from the R DataFrame. The key step is to expose the style formatting with `format()` and then use its methods to style the resulting HTML

Date	High	Low	Open	Close	Volume	Adj Close
01/02/2020	9.420000	9.190000	9.290000	9.420000	43425700	9.262475
01/03/2020	9.370000	9.150000	9.310000	9.210000	45040800	9.055987
01/06/2020	9.170000	9.060000	9.100000	9.160000	43372300	9.006823
01/07/2020	9.250000	9.120000	9.200000	9.250000	44984100	9.095318
01/08/2020	9.300000	9.170000	9.230000	9.250000	45994900	9.095318
01/09/2020	9.310000	9.180000	9.300000	9.260000	51817400	9.105151
01/10/2020	9.360000	9.250000	9.270000	9.250000	39796300	9.095318

Fig. 5.2 HTML-highlighted items in DataFrame

Date	High	Low	Open	Close	Volume	Adj Close
01/02/2020	9.420000	9.190000	9.290000	9.420000	43425700	9.262475
01/03/2020	9.370000	9.150000	9.310000	9.210000	45040800	9.055987
01/06/2020	9.170000	9.060000	9.100000	9.160000	43372300	9.006823
01/07/2020	9.250000	9.120000	9.200000	9.250000	44984100	9.095318
01/08/2020	9.300000	9.170000	9.230000	9.250000	45994900	9.095318
01/09/2020	9.310000	9.180000	9.300000	9.260000	51817400	9.105151
01/10/2020	9.360000	9.250000	9.270000	9.250000	39796300	9.095318

Fig. 5.3 Custom color gradients for HTML DataFrame rendering

Date	High	Low	Open	Close	Volume	Adj Close
01/02/2020	9.420000	9.190000	9.290000	9.420000	43425700	9.262475
01/03/2020	9.370000	9.150000	9.310000	9.210000	45040800	9.055987
01/06/2020	9.170000	9.060000	9.100000	9.160000	43372300	9.006823
01/07/2020	9.250000	9.120000	9.200000	9.250000	44984100	9.095318
01/08/2020	9.300000	9.170000	9.230000	9.250000	45994900	9.095318
01/09/2020	9.310000	9.180000	9.300000	9.260000	51817400	9.105151
01/10/2020	9.360000	9.250000	9.270000	9.250000	39796300	9.095318

Fig. 5.4 Customized background barcharts for DataFrame

table in Jupyter Notebook. The following changes the color-wise visual gradient of the Volume column as in Fig. 5.3.

```
>>> (df.style.format(dict(Date='{:%m/%d/%Y}'))
...    .hide_index()
...    .background_gradient(subset='Volume',cmap='Blues')
... )
<pandas.io.formats.style.Styler object at 0x7f9376a8a0d0>
```

Background barcharts can also be embedded in the table representation in Jupyter Notebook, as in the following (see Fig. 5.4),

```
>>> (df.style.format(dict(Date='{:%m/%d/%Y}'))
...    .hide_index()
...    .bar('Volume',color='lightblue',align='zero')
... )
<pandas.io.formats.style.Styler object at 0x7f9374711f70>
```

5.11 Multi-index

We encountered the Pandas `MultiIndex` when using `groupby` with multiple columns, but these can be created separately:

```
>>> idx = pd.MultiIndex.from_product([['a','b'],[1,2,3]])
>>> data = 10*np.arange(6).reshape(6,1)
>>> df = pd.DataFrame(data=data,index=idx,columns=['A'])
>>> df
       A
a 1    0
  2   10
  3   20
b 1   30
  2   40
  3   50
```

which is more compact than the following,

```
>>> df.reset_index()
   level_0  level_1    A
0        a        1    0
1        a        2   10
2        a        3   20
3        b        1   30
4        b        2   40
5        b        3   50
```

But recall that we did not give the index a name when we created it, which explains the uninformative headers, `level_0` and `level_1`. We can swap the two levels of the index in the DataFrame,

```
>>> df.swaplevel()
       A
1 a    0
2 a   10
3 a   20
1 b   30
2 b   40
3 b   50
```

The `pd.IndexSlice` makes it much easier to index the `DataFrame` using the `loc` accessor,

```
>>> ixs = pd.IndexSlice
>>> df.loc[ixs['a',:],:]
       A
a 1    0
  2   10
  3   20
>>> df.loc[ixs[:,2],:]
       A
a 2   10
b 2   40
```

Note that there can be many more levels to a multi-index. These can also go in the column index,

```
>>> rx = pd.MultiIndex.from_product([[['a','b'],[1,2,3]])
>>> cx = pd.MultiIndex.from_product([[['A','B','C'],[2,3]])
>>> data=[[2, 3, 9, 3, 4, 1],
...       [9, 5, 9, 7, 2, 1],
...       [9, 4, 4, 3, 2, 1],
...       [1, 0, 4, 5, 5, 5],
...       [5, 8, 1, 6, 1, 7],
...       [0, 8, 9, 2, 1, 9]]
>>> df = pd.DataFrame(index=rx,columns=cx,data=data)
>>> df
      A     B     C
      2  3  2  3  2  3
a 1   2  3  9  3  4  1
  2   9  5  9  7  2  1
  3   9  4  4  3  2  1
b 1   1  0  4  5  5  5
  2   5  8  1  6  1  7
  3   0  8  9  2  1  9
```

You can use pd.IndexSlice for both the columns and rows,

```
>>> df.loc[ixs['a',:],ixs['A',:]]=1
>>> df
      A     B     C
      2  3  2  3  2  3
a 1   1  1  9  3  4  1
  2   1  1  9  7  2  1
  3   1  1  4  3  2  1
b 1   1  0  4  5  5  5
  2   5  8  1  6  1  7
  3   0  8  9  2  1  9
```

It is helpful to add names to the levels of the index,

```
>>> df.index = df.index.set_names(['X','Y'])
>>> df
        A     B     C
        2  3  2  3  2  3
X Y
a 1     1  1  9  3  4  1
  2     1  1  9  7  2  1
  3     1  1  4  3  2  1
b 1     1  0  4  5  5  5
  2     5  8  1  6  1  7
  3     0  8  9  2  1  9
```

Even with these complicated multi-indices on the rows/columns, the groupby method still works, but with a full specification of the particular column as ('B',2),

```
>>> df.groupby(('B',2)).sum()
        A     B  C
        2  3  3  2  3
```

```
(B, 2)
1        5    8    6    1    7
4        2    1    8    7    6
9        2   10   12    7   11
```

To understand how this works, take the column slice and examine its unique elements. This explains the values in the resulting row index of the output.

```
>>> df.loc[:,('B',2)].unique()
array([9, 4, 1])
```

Now, we have to examine the partitions that are created in the DataFrame by each of these values such as:

```
>>> df.groupby(('B',2)).get_group(4)
     A      B      C
     2   3   2   3   2   3
X Y
a 3  1   1   4   3   2   1
b 1  1   0   4   5   5   5
```

and then summing over these groups produces the final output. You can also use the apply function on the group to calculate non-scalar output. For example, to subtract the minimum of each element in the group we can do the following,

```
>>> df.groupby(('B',2)).apply(lambda i:i-i.min())
     A      B      C
     2   3   2   3   2   3
X Y
a 1  1   0   0   1   3   0
  2  1   0   0   5   1   0
  3  0   1   0   0   0   0
b 1  0   0   0   2   3   4
  2  0   0   0   0   0   0
  3  0   7   0   0   0   8
```

5.12 Pipes

Pandas implements method-chaining with the pipe function. Even though this is un-Pythonic, it is easier than composing functions together that manipulate dataframes from end-to-end.

```
>>> df = pd.DataFrame(index=['a','b','c'],
...                   columns=['A','B','C'],
...                   data = np.arange(3*3).reshape(3,3))
>>> df.pipe(lambda i:i*10).pipe(lambda i:3*i)
     A    B    C
a    0   30   60
b   90  120  150
c  180  210  240
```

Suppose we need to find the cases for which the sum of the columns is an odd number. We can create an intermediate throw-away variable t using `assign` and then extract the corresponding section of the `DataFrame` as in the following,

```
>>> df.assign(t=df.A+df.B+df.C).query('t%2==1').drop('t',axis=1)
   A  B  C
a  0  1  2
c  6  7  8
```

The `assign` method takes either a function whose argument is the `DataFrame` itself or the named `DataFrame`. The `query` method then filters the intermediate result according to the oddness of t and the final step is to remove the t variable that we no longer need in the output.

5.13 Data Files and Databases

Pandas has powerful I/O utilities for manipulating Excel and CSV spreadsheets.

```
>>> df.to_excel('this_excel.file.xls')
```

You will find that the given spreadsheet has the dates formatted according to Excel's internal date representation.

If you have PyTables installed, you can write to HDFStore. You can also manipulate HDF5 directly from PyTables.

```
>>> df.to_hdf('filename.h5','keyname')
```

You can later read this using

```
>>> dg=pd.read_hdf('filename.h5','keyname')
```

to get your data back. You can create a SQLite database right away because SQLite is included with Python itself.

```
>>> import sqlite3
>>> cnx = sqlite3.connect(':memory:')
>>> df = pd.DataFrame(index=['a','b','c'],
...                   columns=['A','B','C'],
...                   data = np.arange(3*3).reshape(3,3))
>>> df.to_sql('TableName',cnx)
```

Now, we can reload from the database using the usual relational algebra

```
>>> from pandas.io import sql
>>> p2 = sql.read_sql_query('select * from TableName', cnx)
>>> p2
  index  A  B  C
0     a  0  1  2
1     b  3  4  5
2     c  6  7  8
```

5.14   Customizing Pandas

Since Pandas 0.23, we have `extensions.register_dataframe_accessor`, which allows for easy extension of Pandas Dataframes/Series without subclassing.

```
>>> df = pd.DataFrame(index=['a','b','c'],
...                    columns=['A','B','C','D'],
...                    data = np.arange(3*4).reshape(3,4))
>>> df
   A  B   C   D
a  0  1   2   3
b  4  5   6   7
c  8  9  10  11
```

The following code defines a custom accessor that behaves as if it is a native `DataFrame` method.

```
>>> @pd.api.extensions.register_dataframe_accessor('custom')
... class CustomAccess:
...     def __init__(self,df): # receives DataFrame
...         assert 'A' in df.columns # some input validation
...         assert 'B' in df.columns
...         self._df = df
...     @property   # custom attribute
...     def odds(self):
...         'drop all columns that have all even elements'
...         df = self._df
...         return df[df % 2==0].dropna(axis=1,how='all')
...     def avg_odds(self): # custom method
...         'average only odd terms in each column'
...         df = self._df
...         return df[df % 2==1].mean(axis=0)
...
```

Now, with that established, we can use our new method prefixed with the `custom` namespace, as in the following,

```
>>> df.custom.odds # as attribute
   A   C
a  0   2
b  4   6
c  8  10
>>> df.custom.avg_odds() # as method
A    nan
B   5.00
C    nan
D   7.00
dtype: float64
```

Importantly, you can use any word you want besides `custom`. You just have to specify it in the decorator. The analogous `register_series_accessor` does the same thing for `Series` objects and the `register_index_accessor` for `Index` objects.[2]

5.15 Rolling and Filling Operations

Due to Pandas' legacy in quantitative finance, many rolling time-series calculations are easy to compute. Let us load some stock price data.

```
>>> from pandas_datareader import data
>>> df=data.DataReader("F", 'yahoo', '20200101',
↪    '20200130').reset_index()
>>> df.head()
        Date  High   Low  Open  Close       Volume  Adj Close
0 2020-01-02  9.42  9.19  9.29   9.42 43425700.00       9.26
1 2020-01-03  9.37  9.15  9.31   9.21 45040800.00       9.06
2 2020-01-06  9.17  9.06  9.10   9.16 43372300.00       9.01
3 2020-01-07  9.25  9.12  9.20   9.25 44984100.00       9.10
4 2020-01-08  9.30  9.17  9.23   9.25 45994900.00       9.10
```

We can compute the mean over the trailing three elements,

```
>>> df.rolling(3).mean().head(5)
   High   Low  Open  Close       Volume  Adj Close
0   nan   nan   nan   nan          nan       nan
1   nan   nan   nan   nan          nan       nan
2  9.32  9.13  9.23  9.26 43946266.67      9.11
3  9.26  9.11  9.20  9.21 44465733.33      9.05
4  9.24  9.12  9.18  9.22 44783766.67      9.07
```

Note that we only get valid outputs for a fully filled trailing window. Whether or not the endpoints of the window are used in the calculation is determined by the `closed` keyword argument. Besides the default rectangular window, other windows such as Blackman and Hamming are available. The `df.rolling()` function produces a `Rolling` object with methods such as `apply`, `aggregate`, and others. Similar to `rolling`, exponentially weighted windowed calculations can be computed with the `ewm()` method,

```
>>> df.ewm(3).mean()
   High   Low  Open  Close       Volume  Adj Close
0  9.42  9.19  9.29   9.42 43425700.00       9.26
1  9.39  9.17  9.30   9.30 44348614.29       9.14
2  9.30  9.12  9.21   9.24 43926424.32       9.08
3  9.28  9.12  9.21   9.24 44313231.43       9.09
4  9.29  9.14  9.22   9.25 44864456.98       9.09
5  9.29  9.15  9.24   9.25 46979043.75       9.10
6  9.31  9.18  9.25   9.25 44906738.44       9.10
7  9.30  9.16  9.25   9.25 45919910.43       9.09
8  9.31  9.17  9.24   9.26 45113266.20       9.10
```

---

[2]The third-party data-cleaning module `pyjanitor` utilizes this approach extensively.

```
9    9.30 9.18  9.25   9.24 47977202.99        9.09
10   9.30 9.17  9.24   9.22 47020077.94        9.07
11   9.28 9.16  9.23   9.21 45632324.49        9.05
12   9.27 9.14  9.21   9.21 46637216.86        9.05
13   9.26 9.15  9.21   9.20 44926124.49        9.04
14   9.24 9.09  9.19   9.18 52761475.78        9.03
15   9.21 9.06  9.17   9.14 56635156.17        8.98
16   9.14 8.99  9.10   9.07 57676520.00        8.92
17   9.11 8.96  9.06   9.05 64587200.41        8.90
18   9.07 8.93  9.01   9.00 63198880.10        8.89
19   9.01 8.88  8.96   8.96 58089908.44        8.88
```

We have barely scratched the surface of what Pandas is capable of and we have completely ignored its powerful features for managing dates and times. There is much more to learn and the online documentation and tutorials at the main Pandas site are great for learning more.

Chapter 6
Visualizing Data

The following sections discuss visualizing data using key Python modules, but before we get to that, it is important to grasp specific principles of data visualization because these modules can handle the *how* of plotting but not the *what*.

Before building a data visualization, the main thing is to put yourself in the position of the viewer. This part is easy to get wrong because you (as the author) typically have a larger visual vocabulary than the viewer, and, combined with your familiarity with the data, makes you prone to overlook many points of confusion from the viewer's perspective.

To make this worse, human perception is mainly a figment of the imagination, driven to compensate for the limitations of our physical visual hardware. This means that it is not just beauty that is in the eye of the beholder, but so is your data visualization! Good data science strives to communicate dispassionate facts but our emotional response to visuals is pre-cognitive, meaning that we have already responded to the presentation far before we are consciously aware of doing so. Viewers later justify their initial emotional impressions of the presentation after the fact, rather than seriously coming to a reasoned conclusion. Thus the visualization author must constantly be on guard for such problems.

Nothing we are discussing here is *new* and the visual presentation of quantitative information is a well-established field, worthy of study in its own right. At a minimum, you should be familiar with the perceptual properties of data graphics as a hierarchy in terms of communication accuracy. At the top (i.e., most accurate) is using position to represent data (i.e., scatter plots) because our human visual cognition is very good at picking out clusters of points in the distance. Thus, whenever possible, strive to use position to communicate the most important aspect of your visual. Next in order is length (i.e., barcharts) because we can easily distinguish differences in length, insofar as everything else in the graphic is aligned. After length is angle, which explains why cockpits in airplanes have gauges that all point in the same direction, allowing the pilot to instantly detect when a gauge is amiss because of the angular deflection of the needle. Next is area and then volume,

© The Editor(s) (if applicable) and The Author(s), under exclusive license
to Springer Nature Switzerland AG 2021
J. Unpingco, *Python Programming for Data Analysis*,
https://doi.org/10.1007/978-3-030-68952-0_6

and then density. It is difficult for us to detect whether two circles have the same area unless one is significantly larger than the other. Volume is the worst as it is strongly influenced by shape. If you have every tried to pour a jug of water into a square pan, then you know how hard that is to judge. Color is also hard because color comes with strong emotional baggage that can be distracting, notwithstanding issues of color-blindness.

Thus, the protocol is to figure out what the main message of the data visualization is and then to use the graphical hierarchy to encode that message using position, whenever possible. Collateral messages are then relegated to items further down the hierarchy like angle or color. Then, remove everything else from the visualization that does not contribute to your messages, including default colors, lines, axes, text, or anything else that gets in the way. Keep in mind that the Python visualization modules come with their own defaults, which may not be conducive to your *particular* desired message.

6.1   Matplotlib

Matplotlib is the primary visualization tool for scientific graphics in Python. Like all great open-source projects, it originated to satisfy a personal need. At the time of its inception, John Hunter primarily used Matlab for scientific visualization, but as he began to integrate data from disparate sources using Python, he realized he needed a Python solution for visualization, so he wrote the initial version of Matplotlib. Since those early years, Matplotlib has displaced the other competing methods for two-dimensional scientific visualization and today is a very actively maintained project, even without John Hunter, who sadly passed away in 2012. Further, other projects like seaborn leverage Matplotlib primitives for specialized plotting. In this way, Matplotlib has soaked into the infrastructure of visualization for the Scientific Python community. The key and enduring strength of Matplotlib is its completeness—just about any kind of two-dimensional scientific plot you can think of can be generated, in full publication quality, using Matplotlib, as can be seen with the *John Hunter Excellence in Plotting Contest*.

There are two main conceptual pieces for Matplotlib: the canvas and artists. The canvas can be thought of as the target for the visualization and the artists *draw* on the canvas. To create a canvas in Matplotlib, you can use the plt.figure function shown below. Then, the plt.plot function puts Line2D artists onto the canvas that draw Fig. 6.1 and are returned as output in a Python list. Later, we will manipulate these returned artists.

```
>>> import numpy as np
>>> import matplotlib.pyplot as plt
>>> plt.figure() # setup figure
<Figure size 640x480 with 0 Axes>
>>> x = np.arange(10) # Create some data
>>> y = x**2
>>> plt.plot(x,y)
```

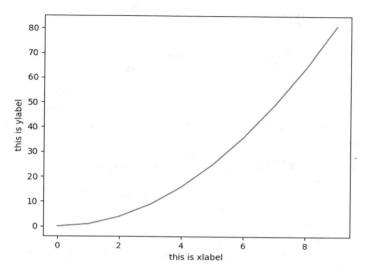

Fig. 6.1 Basic Matplotlib plot

```
[<matplotlib.lines.Line2D object at 0x7f9373184850>]
>>> plt.xlabel('This is the xlabel') # apply labels
Text(0.5, 0, 'This is the xlabel')
>>> plt.ylabel('This is the ylabel')
Text(0, 0.5, 'This is the ylabel')
>>> plt.show() # show figure
```

Note that if you execute the above in a plain Python interpreter, the process will freeze (or, *block*) at the last line. This is because Matplotlib is preoccupied with rendering the GUI window, which is the ultimate target of the resulting graphic. The `plt.show` function triggers the artists to render upon the canvas. The reason this comes last is so that the artists can be mustered to render all at once.

Although using these functions is the traditional way to use Matplotlib, the object-oriented interface is more organized and compact. The following re-does the above using the object-oriented style:

```
>>> from matplotlib.pylab import subplots
>>> fig,ax = subplots()
>>> ax.plot(x,y)
>>> ax.set_xlabel('this is xlabel')
>>> ax.set_ylabel('this is ylabel')
```

Note that the key step is to use the `subplots` function to generate objects for the figure window and the axes. Then, plotting commands are thereby attached to the respective `ax` object. This makes it easier to keep track of multiple visualizations overlaid on the same `ax`.

6.1.1   Setting Defaults

You can set the defaults for plotting using `rcParams` dictionary

```
>>> import matplotlib
>>> matplotlib.rcParams['lines.linewidth']=2.0 # all plots with
↪   have this linewidth
```

Alternatively, you could have set the `linewidth` on a per-line basis using keyword arguments,

```
plot(arange(10),linewidth=2.0) # using linewidth keyword
```

6.1.2   Legends

Legends identify the lines on the plot (see Fig. 6.2) and `loc` indicates the position of the legend.

```
>>> fig,ax=subplots()
>>> ax.plot(x,y,x,2*y,'or--')
>>> ax.legend(('one','two'),loc='best')
```

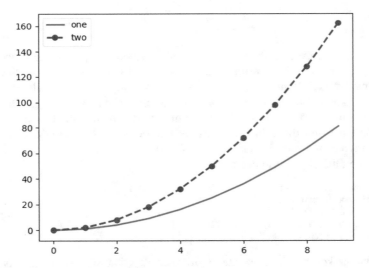

Fig. 6.2 Multiple lines on the same axes with a legend

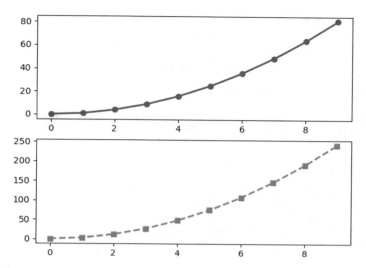

Fig. 6.3 Multiple subplots in the same figure

6.1.3   Subplots

The same `subplots` function permits multiple subplots (see Fig. 6.3) with each
axis indexed like a Numpy array,

```
>>> fig,axs = subplots(2,1) # 2-rows, 1-column
>>> axs[0].plot(x,y,'r-o')
>>> axs[1].plot(x,y*3,'g--s')
```

6.1.4   Spines

The rectangle that contains the plots has four so-called *spines*. These are managed
with the `spines` object. In the example below, note that the `axs` contains a slice-
able array of individual axes objects (see Fig. 6.4). The subplot on the top left
corresponds to `axs[0,0]`. For this subplot, the spine on the right is made invisible
by setting its color to `'none'`. Note that for Matplotlib, the string `'none'` is
handled differently than the usual Python `None`. The spine at the bottom is moved
to the `center` using `set_position` and the positions of the ticks are assigned with
`set_ticks_position`.

```
>>> fig,axs = subplots(2,2)
>>> x = np.linspace(-np.pi,np.pi,100)
>>> y = 2*np.sin(x)
>>> ax = axs[0,0]
>>> ax.set_title('centered spines')
```

```
>>> ax.plot(x,y)
>>> ax.spines['left'].set_position('center')
>>> ax.spines['right'].set_color('none')
>>> ax.spines['bottom'].set_position('center')
>>> ax.spines['top'].set_color('none')
>>> ax.xaxis.set_ticks_position('bottom')
>>> ax.yaxis.set_ticks_position('left')
```

The next subplot on the bottom left, axes[1,0] has the bottom spine moved to
the 'zero' position of the data.

```
>>> ax = axs[1,0]
>>> ax.set_title('zeroed spines')
>>> ax.plot(x,y)
>>> ax.spines['left'].set_position('zero')
>>> ax.spines['right'].set_color('none')
>>> ax.spines['bottom'].set_position('zero')
>>> ax.spines['top'].set_color('none')
>>> ax.xaxis.set_ticks_position('bottom')
>>> ax.yaxis.set_ticks_position('left')
```

The next subplot on the top right, axes[0,1] has the bottom spine moved to the
0.1 position of the axes coordinate system with the left spine at the 0.6 position
of the axes coordinate system (more about coordinate systems later).

```
>>> ax = axs[0,1]
>>> ax.set_title('spines at axes (0.6, 0.1)')
>>> ax.plot(x,y)
>>> ax.spines['left'].set_position(('axes',0.6))
>>> ax.spines['right'].set_color('none')
>>> ax.spines['bottom'].set_position(('axes',0.1))
>>> ax.spines['top'].set_color('none')
>>> ax.xaxis.set_ticks_position('bottom')
>>> ax.yaxis.set_ticks_position('left')
```

6.1.5   Sharing Axes

The subplots function also allows sharing of individual axes between plots using
sharex and sharey (see Fig. 6.5) which is particularly helpful for aligning time-
series plots.

```
>>> fig, axs = subplots(3,1,sharex=True,sharey=True)
>>> t = np.arange(0.0, 2.0, 0.01)
>>> s1 = np.sin(2*np.pi*t)
>>> s2 = np.exp(-t)
>>> s3 = s1*s2
>>> axs[0].plot(t,s1)
>>> axs[1].plot(t,s2)
>>> axs[2].plot(t,s3)
>>> ax.set_xlabel('x-coordinate')
```

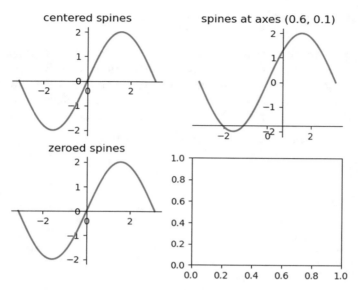

Fig. 6.4 Spines refer to the frame edges and can be moved within the figure

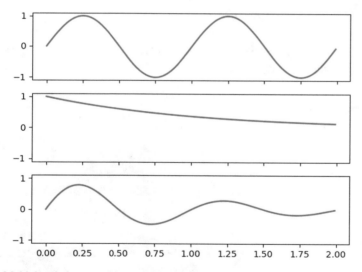

Fig. 6.5 Multiple subplots can share a single x-axis

6.1.6   3D Surfaces

Matplotlib is primarily a two-dimensional plotting package, but it has some limited three-dimensional capabilities that are primarily based on projection mechanisms. The main machinery for this is in the `mplot3d` submodule. The following code draws Fig. 6.6. The `Axes3D` object has the `plot_surface` method for drawing the three-dimensional graphic while the `cm` module has the colormaps.

```
>>> from mpl_toolkits.mplot3d import Axes3D
>>> from matplotlib import cm
>>> fig = plt.figure()
>>> ax = Axes3D(fig)
>>> X = np.arange(-5, 5, 0.25)
>>> Y = np.arange(-5, 5, 0.25)
>>> X, Y = np.meshgrid(X, Y)
>>> R = np.sqrt(X**2 + Y**2)
>>> Z = np.sin(R)
>>> ax.plot_surface(X, Y, Z, rstride=1, cstride=1, cmap=cm.jet)
<mpl_toolkits.mplot3d.art3d.Poly3DCollection object at
0x7f9372d6b5b0>
```

6.1.7   Using Patch Primitives

You can plot circles, polygons, etc. using primitives available in the `matplotlib.patches` module (see Fig. 6.7).

Fig. 6.6 Three-dimensional surface plot

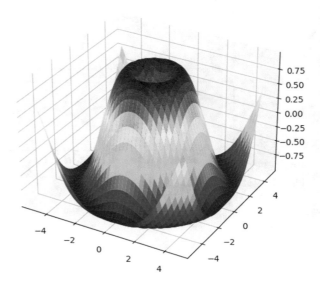

Fig. 6.7 Matplotlib *patches* draw graphic primitives onto the canvas

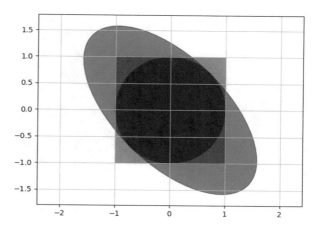

```
>>> from matplotlib.patches import Circle, Rectangle, Ellipse
>>> fig,ax = subplots()
>>> ax.add_patch(Circle((0,0), 1,color='g'))
>>> ax.add_patch(Rectangle((-1,-1),
...             width = 2, height = 2,
...             color='r',
...             alpha=0.5)) # transparency
>>> ax.add_patch(Ellipse((0,0),
...             width = 2, height = 4, angle = 45,
...             color='b',
...             alpha=0.5))   # transparency
>>> ax.axis('equal')
(-1.795861705, 1.795861705, -1.795861705, 1.795861705)
>>> ax.grid(True)
```

You can use cross-hatches instead of colors as well (see Fig. 6.8).

```
>>> fig,ax = subplots()
>>> ax.add_patch(Circle((0,0),
...                   radius=1,
...                   facecolor='w',
...                   hatch='x'))
>>> ax.grid(True)
>>> ax.set_title('Using cross-hatches',fontsize=18)
```

6.1.8 Patches in 3D

You can also add patches to 3D axes. The three-dimensional axis is created by passing the `subplot_kw` keyword argument the `{'projection':'3d'}` dictionary. The `Circle` patches are created in the usual way and then added to this axis. The `pathpatch_2d_to_3d` method from the `art3d` module changes the viewer's

Fig. 6.8 Cross-hatching
instead of colors

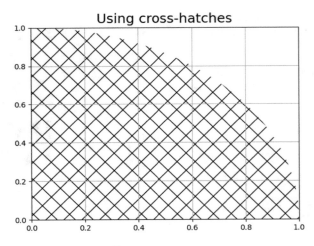

Fig. 6.9 Matplotlib patch
primitives can be drawn in
three dimensions also

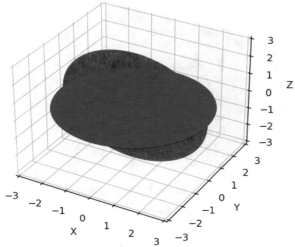

perspective to each patch along the indicated direction. This is what creates the
three-dimensional effect (see Fig. 6.9).

```
>>> import mpl_toolkits.mplot3d.art3d as art3d
>>> fig, ax  = subplots(subplot_kw={'projection':'3d'})
>>> c = Circle((0,0),radius=3,color='r')
>>> d = Circle((0,0),radius=3,color='b')
>>> ax.add_patch(c)
>>> ax.add_patch(d)
>>> art3d.pathpatch_2d_to_3d(c,z=0,zdir='y')
>>> art3d.pathpatch_2d_to_3d(d,z=0,zdir='z')
```

Two-dimensional graphs can also be stacked in three dimensions (see Fig. 6.10). The
paths of the individual closed polygons of the plots are extracted using get_paths

Fig. 6.10 Custom patch primitives can be created from other Matplotlib renderings

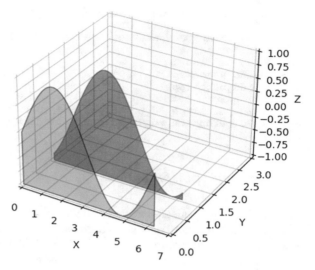

Fig. 6.11 Graphical elements in different coordinate systems appear on the same figure

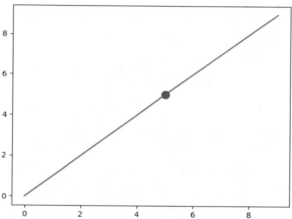

and then used to create the `PathPatch` objects are added to the axis. As before, the final step is to set the perspective view on each patch using `pathpatch_2d_to_3d`.

```
>>> from numpy import pi, arange, linspace, sin, cos
>>> from matplotlib.patches import PathPatch
>>> x = linspace(0,2*pi,100)
>>> # create polygons from graphs
>>> fig, ax = subplots()
>>> p1=ax.fill_between(x,sin(x),-1)
>>> p2=ax.fill_between(x,sin(x-pi/3),-1)
>>> path1=p1.get_paths()[0] # get closed polygon for p1
>>> path2=p2.get_paths()[0] # get closed polygon for p2
>>> ax.set_title('setting up patches from 2D graph')
>>> fig, ax   = subplots(subplot_kw={'projection':'3d'})
```

```
>>> pp1 = PathPatch(path1,alpha=0.5) # need to assign this for
↪    later
>>> pp2 = PathPatch(path2,color='r',alpha=0.5) # need to assign
↪    this for later
>>> # add patches
>>> ax.add_patch(pp1)
>>> ax.add_patch(pp2)
>>> # transform patches
>>> art3d.pathpatch_2d_to_3d(pp1,z=0,zdir='y')
>>> art3d.pathpatch_2d_to_3d(pp2,z=1,zdir='y')
```

6.1.9   Using Transformations

One of Matplotlib's underappreciated strengths is how it manages multiple coordinate systems as it creates data visualizations. These coordinate systems provide the scale mappings between the data and the space of the rendered visualization. The data coordinate system is the coordinate system of the data points whereas the display coordinate system is the system of the displayed figure. The ax.transData method converts the data coordinates (5, 5) into coordinates in the display coordinate system (see Fig. 6.11).

```
>>> fig,ax = subplots()
>>> # line in data coordinates
>>> ax.plot(np.arange(10), np.arange(10))
>>> # marks the middle in data coordinates
>>> ax.plot(5,5,'o',markersize=10,color='r')
>>> # show the same point but in display coordinates
>>> print(ax.transData.transform((5,5)))
[2560.   1900.8]
```

If we create axes with different sizes using the figsize keyword argument, then the transData method returns different coordinates for each axis, even though they refer to the same point in data coordinates.

```
>>> fig,ax = subplots(figsize=(10,4))
>>> ax.transData.transform((5,5))
array([4000., 1584.])
>>> fig,ax = subplots(figsize=(5,4))
>>> ax.transData.transform((5,5))
array([2000., 1584.])
```

Furthermore, if you plot this in a GUI window (not in the Jupyter notebook) and resize the figure using the mouse and do this again, you also get different coordinates depending on how the window was resized. As a practical matter, you rarely work in display coordinates. You can go back to data coordinates using ax.transData.inverted().transform.

The axes coordinate system is the unit box that contains the axes. The transAxes method maps into this coordinate system. For example, see Fig. 6.12 where the transAxes method is used for the transform keyword argument,

Fig. 6.12 Elements can be
fixed to positions in the figure
regardless of the data
coordinates using the axes
coordinate system

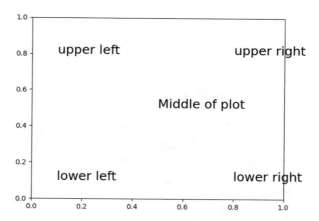

```
>>> from matplotlib.pylab import gca
>>> fig,ax = subplots()
>>> ax.text(0.5,0.5,
...         'Middle of plot',
...         transform = ax.transAxes,
...         fontsize=18)
>>> ax.text(0.1,0.1,
...         'lower left',
...         transform = ax.transAxes,
...         fontsize=18)
>>> ax.text(0.8,0.8,
...         'upper right',
...         transform = ax.transAxes,
...         fontsize=18)
>>> ax.text(0.1,0.8,
...         'upper left',
...         transform = ax.transAxes,
...         fontsize=18)
>>> ax.text(0.8,0.1,
...         'lower right',
...         transform = ax.transAxes,
...         fontsize=18)
```

This can be useful for annotating plots consistently irrespective of the data. You can
combine this with the patches to create a mix of data coordinate and axes coordinate
items in your graphs as in Fig. 6.13, from the following code:

```
>>> fig,ax = subplots()
>>> x = linspace(0,2*pi,100)
>>> ax.plot(x,sin(x) )
>>> ax.add_patch(Rectangle((0.1,0.5),
...         width = 0.5,
...         height = 0.2,
...         color='r',
...         alpha=0.3,
...         transform = ax.transAxes))
>>> ax.axis('equal')
```

```
(-0.3141592653589793, 6.5973445725385655,
-1.0998615404412626, 1.0998615404412626)
```

6.1.10   Annotating Plots with Text

Matplotlib implements a unified class for text which means that you can manipulate text the same way anywhere it appears in a visualization. Adding text is straightforward with ax.text (see Fig. 6.14):

```
>>> fig,ax = subplots()
>>> x = linspace(0,2*pi,100)
>>> ax.plot(x,sin(x) )
>>> ax.text(pi/2,1,'max',fontsize=18)
>>> ax.text(3*pi/2,-1.1,'min',fontsize=18)
>>> ax.text(pi,0,'zero',fontsize=18)
>>> ax.axis((0,2*pi,-1.25,1.25))
(0.0, 6.283185307179586, -1.25, 1.25)
```

We can also use bounding boxes to surround text as in the following Fig. 6.15 by using the bbox keyword argument,

```
>>> fig,ax = subplots()
>>> x = linspace(0,2*pi,100)
>>> ax.plot(x,sin(x))
>>> ax.text(pi/2,1-0.5,'max',
...             fontsize=18,
...             bbox = {'boxstyle':'square','facecolor':'yellow'})
```

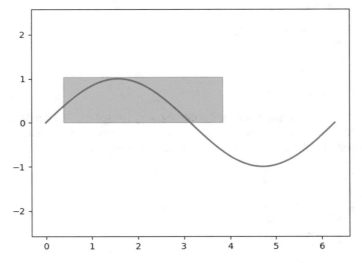

Fig. 6.13 Matplotlib patch primitives can be used in different coordinate systems

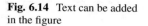

Fig. 6.14 Text can be added
in the figure

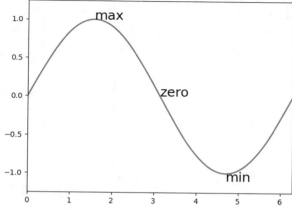

Fig. 6.15 Bounding boxes
can be added to text

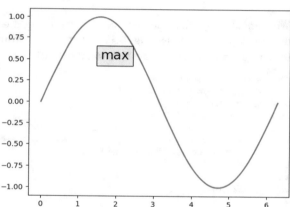

6.1.11 *Annotating Plots with Arrows*

Plots can be annotated with arrows using `ax.annotate` as in the following
(Fig. 6.16):

```
>>> fig,ax = subplots()
>>> x = linspace(0,2*pi,100)
>>> ax.plot(x,sin(x))
>>> ax.annotate('max',
...     xy=(pi/2,1), # where to put arrow endpoint
...     xytext=(pi/2,0.3), # text position in data coordinates
...     arrowprops={'facecolor':'black','shrink':0.05},
...     fontsize=18,
...     )
>>> ax.annotate('min',
...     xy=(3/2.*pi,-1), # where to put arrow endpoint
...     xytext=(3*pi/2.,-0.3), # text position in data coordinates
...     arrowprops={'facecolor':'black','shrink':0.05},
...     fontsize=18,
...     )
```

Fig. 6.16 Arrows call
attention to points on the
graph

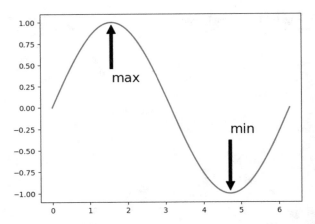

We can also specify the text coordinate using `textcoords` system, as we
discussed previously (see Fig. 6.17) where the coordinate system is specified with
the string `'axes fraction'` instead of using `ax.transAxes`.

```
>>> fig,ax = subplots()
>>> x = linspace(0,2*pi,100)
>>> ax.plot(x,sin(x) )
>>> ax.annotate('max',
...     xy=(pi/2,1),        # where to put arrow endpoint
...     xytext=(0.3,0.8),   # text position in axes coordinates
...     textcoords='axes fraction',
...     arrowprops={'facecolor':'black',
...                 'shrink':0.05},
...     fontsize=18,
...     )
>>> ax.annotate('min',
...     xy=(3/2.*pi,-1),    # where to put arrow endpoint
...     xytext=(0.8,0.2),   # text position in data coordinates
...     textcoords='axes fraction',
...     arrowprops={'facecolor':'black',
...                 'shrink':0.05,
...                 'width':10,
...                 'headwidth':20,
...                 'headlength':6},
...     fontsize=18,
...     )
```

Sometimes, you just want the arrow and *not* the text as in Fig. 6.18, where `xytext`
coordinates are for the tail of the arrow and `connectionstyle` specifies the
curve of the arc.

```
>>> fig,ax = subplots()
>>> x = linspace(0,2*pi,100)
>>> ax.set_title('Arrow without text',fontsize=18)
>>> ax.annotate("", # leave the text-string argument empty
...         xy=(0.2, 0.2), xycoords='data',
...         xytext=(0.8, 0.8), textcoords='data',
```

Fig. 6.17 The representation of arrows can be carefully detailed

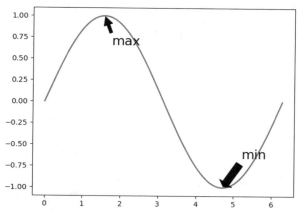

Fig. 6.18 Arrow without the text

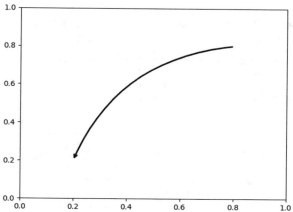

```
...        arrowprops=dict(arrowstyle="->",
...                        connectionstyle="arc3,rad=0.3",
...                        linewidth=2.0),
...        )
```

As arrows are important for pointing to graphical elements or for representing complicated vector fields (e.g., wind-speeds and directions), Matplotlib provides many customization options.

6.1.12 Embedding Scalable and Non-scalable Subplots

Embedded subplots can scale or not scale with the rest of the plotted data. The following embedded circles in Fig. 6.19 will not change even if other data are plotted within the same figure via AnchoredDrawingArea. Note that we artists

Fig. 6.19 Embedded
subplots can be independent
of scaling data in the same
figure

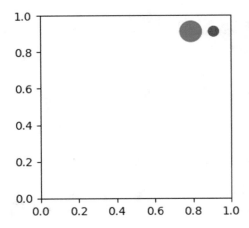

are added to the anchored drawing area, their dimensions are in pixel coordinates,
even if the `transform` keyword argument is set for the artist.

```
>>> from mpl_toolkits.axes_grid1.anchored_artists import
AnchoredDrawingArea
>>> fig,ax = subplots()
>>> fig.set_size_inches(3,3)
>>> ada = AnchoredDrawingArea(40, 20, 0, 0,
...                           loc=1, pad=0.,
...                           frameon=False)
>>> p1 = Circle((10, 10), 10)
>>> ada.drawing_area.add_artist(p1)
>>> p2 = Circle((30, 10), 5, fc="r")
>>> ada.drawing_area.add_artist(p2)
>>> ax.add_artist(ada)
<mpl_toolkits.axes_grid1.anchored_artists.AnchoredDrawingArea
object at 0x7f939833e760>
```

This is different from `AnchoredAuxTransformBox` which will facilitate scaling with data coordinates in Fig. 6.20 where the embedded ellipse scales with `ax`,

```
>>> from matplotlib.patches import Ellipse
>>> from mpl_toolkits.axes_grid1.anchored_artists import
AnchoredAuxTransformBox

>>> fig,ax = subplots()
>>> box = AnchoredAuxTransformBox(ax.transData, loc=2)
>>> # in data coordinates
>>> el = Ellipse((0,0),
...        width=0.1,
...        height=0.4,
...        angle=30)
>>> box.drawing_area.add_artist(el)
>>> ax.add_artist(box)
<mpl_toolkits.axes_grid1.anchored_artists.AnchoredAuxTransformBox
object at 0x7f9372f5dbe0>
```

Adding plots to this *will* leave the circles unchanged.

Fig. 6.20 As opposed to
Fig. 6.19, this subplot *will*
change with the scaling of
other data graphics in this
figure

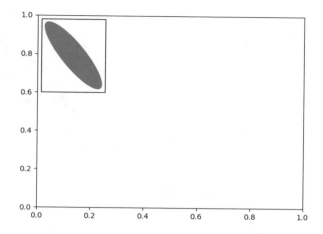

6.1.13   Animations

Matplotlib provides flexible animations that utilize all of the other graphical
primitives. One way to create animations is to generate a sequences of artists in
an iterable and then animate them (see Fig. 6.21).

```
>>> import matplotlib.animation as animation
>>> fig,ax = subplots()
>>> x = np.arange(10)
>>> frames = [ax.plot(x,x,x[i],x[i],'ro',ms=5+i*10)
...              for i in range(10)]
>>> # You must assign in the next line!
>>> g=animation.ArtistAnimation(fig,frames,interval=50)
```

This works fine for relatively few frames, but you can also dynamically create
frames on the fly (see Fig. 6.22):

```
>>> import matplotlib.animation as animation
>>> x = np.arange(10)
>>> linewidths =[10,20,30]
>>> fig = plt.figure()
>>> ax = fig.add_subplot(111)
>>> line, = ax.plot(x,x,'-ro',ms=20,linewidth=5.0)
>>> def update(data):
...     line.set_linewidth(data)
...     return (line,)
...
>>> ani = animation.FuncAnimation(fig, update, x, interval=500)
>>> plt.show()
```

To get these to animate in a Jupyter notebook you have to convert them to
corresponding JavaScript animations using to_jshtml. For example, in one cell,
do the following (see Fig. 6.23):

```
>>> fig,ax = subplots()
>>> frames = [ax.plot(x,x,x[i],x[i],'ro',ms=5+i*10)
```

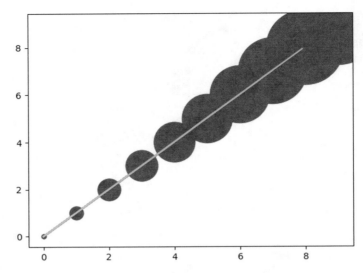

Fig. 6.21 Matplotlib animations utilize the other graphical primitives

Fig. 6.22 Animations can be generated dynamically with functions

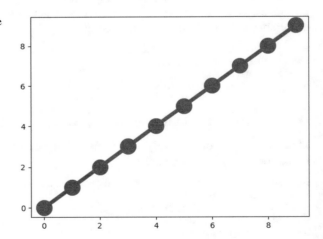

```
...                    for i in range(10)]
>>> g=animation.ArtistAnimation(fig,frames,interval=50)
```

and in the next cell, do the following:

```
from IPython.display import HTML
HTML(g.to_jshtml())
```

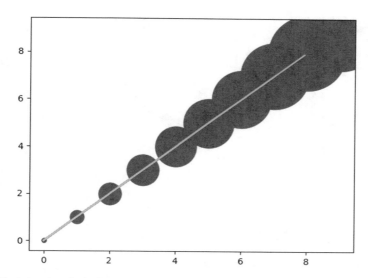

Fig. 6.23 Animations in the Jupyter notebook can be played in the browser using `to_jshtml`

6.1.14 Using Paths Directly

Further low-level access to plotting is available in Matplotlib by using paths, which
you can think of as programming a stylus that is drawing on the canvas. For
example, `patches` are made of `paths`. Paths have vertices and corresponding
draw commands. For instance, for Fig. 6.24, this draws a line between two points

```
>>> from matplotlib.path import Path
>>> vertices=[ (0,0),(1,1) ]
>>> codes = [ Path.MOVETO, # move stylus to (0,0)
...            Path.LINETO] # draw line to (1,1)
>>> path = Path(vertices, codes)   #create path
>>> fig, ax = subplots()
>>> # convert path to patch
>>> patch = PathPatch(path,linewidth=10)
>>> ax.add_patch(patch)
>>> ax.set_xlim(-.5,1.5)
(-0.5, 1.5)
>>> ax.set_ylim(-.5,1.5)
(-0.5, 1.5)
>>> plt.show()
```

Paths can also use quadratic and cubic Bezier curves to connect points, as in
Fig. 6.25,

```
>>> vertices=[ (-1,0),(0,1),(1,0),(0,-1),(-1,0) ]
>>> codes = [Path.MOVETO, # move stylus to (0,0)
...           Path.CURVE3, # draw curve
...           Path.CURVE3, # draw curve
...           Path.CURVE3, # draw curve
```

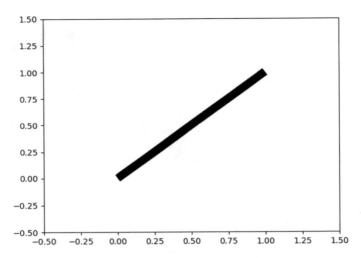

Fig. 6.24 Paths draw on the canvas using specific pen-like instructions

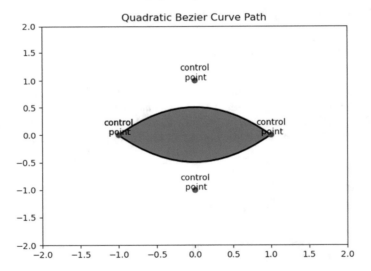

Fig. 6.25 Paths can have Bezier curves

```
...              Path.CURVE3,]
>>> path = Path(vertices, codes)   #create path
>>> fig, ax = subplots()
>>> # convert path to patch
>>> patch = PathPatch(path,linewidth=2)
>>> ax.add_patch(patch)
>>> ax.set_xlim(-2,2)
(-2.0, 2.0)
>>> ax.set_ylim(-2,2)
```

Fig. 6.26 Arrows tangent to
the curve

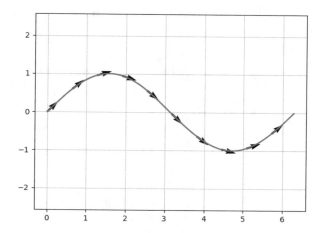

```
(-2.0, 2.0)
>>> ax.set_title('Quadratic Bezier Curve Path')
>>> for i in vertices:
...     _=ax.plot(i[0],i[1],'or'  )# control points
...     _=ax.text(i[0],i[1],'control\n
↪   point',horizontalalignment='center')
...
```

Arrows and plots can be combined into a single figure to show the directional
derivative at points on the curve (see Fig. 6.26),

```
>>> x = np.linspace(0,2*np.pi,100)
>>> y = np.sin(x)
>>> fig, ax = subplots()
>>> ax.plot(x,y)
>>> u = []
>>> # subsample x
>>> x = x[::10]
>>> for i in zip(np.ones(x.shape),np.cos(x)):
...     v=np.array(i)
...     u.append(v/np.sqrt(np.dot(v,v)) )
...
>>> U=np.array(u)
>>> ax.quiver(x,np.sin(x),U[:,0],U[:,1])
>>> ax.grid()
>>> ax.axis('equal')
(-0.3141592653589793, 6.5973445725385655, -1.0998615404412626,
1.0998615404412626)
```

6.1.15 Interacting with Plots Using Sliders

The Matplotlib GUI widget (Qt or other backend) can have callbacks to respond to
key or mouse movements. This makes it easy to add basic interactivity plots in the

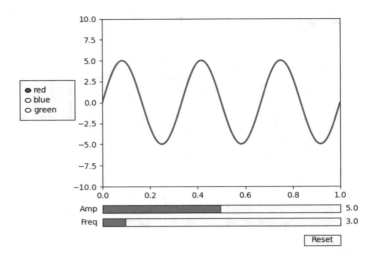

Fig. 6.27   Interactive widgets can be attached to the Matplotlib GUI backend

GUI window. These are also available in `matplotlib.widgets` from within the Jupyter notebook. The widgets are attached to the callback `update` function using `on_changed` or `on_clicked` and changes to the specific widgets trigger the callback (see Fig. 6.27) where the `Slider` , `RadioButtons` and `Button` widgets are attached to callbacks.

```
>>> from matplotlib.widgets import Slider, Button, RadioButtons
>>> from matplotlib.pylab import subplots_adjust, axes
>>> fig, ax = subplots()
>>> subplots_adjust(left=0.25, bottom=0.25)
>>> # setup data
>>> t = arange(0.0, 1.0, 0.001)
>>> a0, f0 = 5, 3
>>> s = a0*sin(2*pi*f0*t)
>>> # draw main plot
>>> l, = ax.plot(t,s, lw=2, color='red')
>>> ax.axis([0, 1, -10, 10])
(0.0, 1.0, -10.0, 10.0)
>>> axcolor = 'lightgoldenrodyellow'
>>> # create axes for widgets
>>> axfreq = axes([0.25, 0.1, 0.65, 0.03], facecolor=axcolor)
>>> axamp  = axes([0.25, 0.15, 0.65, 0.03], facecolor=axcolor)
>>> sfreq = Slider(axfreq, 'Freq', 0.1, 30.0, valinit=f0)
>>> samp = Slider(axamp, 'Amp', 0.1, 10.0, valinit=a0)
>>> def update(val):
...       amp = samp.val
...       freq = sfreq.val
...       l.set_ydata(amp*sin(2*pi*freq*t))
...       draw()
...
>>> # attach callbacks to widgets
>>> sfreq.on_changed(update)
```

```
0
>>> samp.on_changed(update)
0
>>> resetax = axes([0.8, 0.025, 0.1, 0.04])
>>> button = Button(resetax, 'Reset', color=axcolor,
↪  hovercolor='0.975')
>>> def reset(event):
...     sfreq.reset()
...     samp.reset()
...
>>> # attach callback to button
>>> button.on_clicked(reset)
0
>>> rax = axes([0.025, 0.5, 0.15, 0.15], facecolor=axcolor)
>>> radio = RadioButtons(rax, ('red', 'blue', 'green'), active=0)
>>> def colorfunc(label):
...     l.set_color(label)
...     draw()
...
>>> # attach callback to radio buttons
>>> radio.on_clicked(colorfunc)
0
```

6.1.16 Colormaps

Matplotlib provides many useful colormaps. The `imshow` function takes an input array and plots the cells in that array with color corresponding to the value of each entry (see Fig. 6.28).

```
>>> fig, ax  = subplots()
>>> x = np.linspace(-1,1,100)
>>> y = np.linspace(-3,1,100)
>>> ax.imshow(abs(x + y[:,None]*1j)) # use broadcasting
```

Matplotlib colors are organized in the `matplotlib.colors` submodule and the `matplotlib.pylab.cm` module has the colormaps in a single interface. For every named colormap in `cm`, there is a reversed colormap. For example, `cm.Blues` and `cm.Blues_r`. We can use this colormap with `imshow` as in Fig. 6.28 to create Fig. 6.29 with the `cmap` keyword argument.

```
>>> fig, ax  = subplots()
>>> ax.imshow(abs(x + y[:,None]*1j),cmap=cm.Blues)
```

Because colors have a strong perceptual impact for understanding data visually, Matplotlib are organized in terms of cyclic, sequential, diverging, or qualitative, each tailored for specific kinds of numerical, categorical, or nominal data.

Fig. 6.28 Matplotlib
`imshow` displays matrix
elements as colors

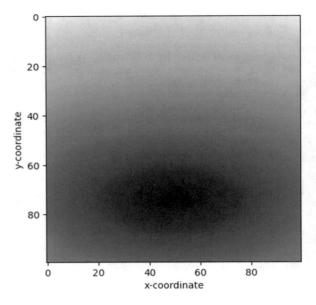

Fig. 6.29 Same as Fig. 6.28
but with the `cm.Blues`
colormap

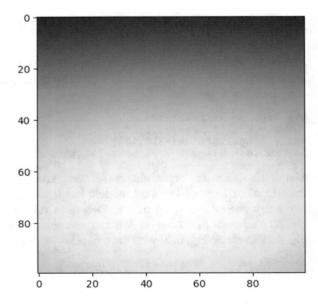

Fig. 6.30 Matplotlib elements' attributes can be discovered and changed using getp and setp

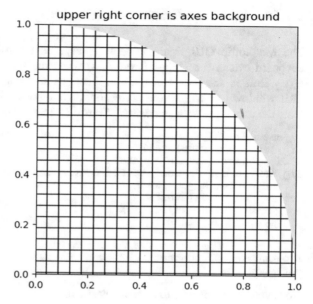

6.1.17 Low-Level Control Using *setp* and *getp*

Because there are so many Matplotlib elements and so many corresponding attributes, the getp and setp functions make it easier to discover and change the attributes of any particular element (see Fig. 6.30).

```
>>> from matplotlib.pylab import setp, getp
>>> fig, ax = subplots()
>>> c = ax.add_patch(Circle((0,0),1,facecolor='w',hatch='-|'))
>>> # set axes background rectangle to blue
>>> setp(ax.get_children()[-1],fc='lightblue')
[None]
>>> ax.set_title('upper right corner is axes background')
>>> ax.set_aspect(1)
>>> fig.show()
```

6.1.18 Interacting with Matplotlib Figures

Matplotlib provides several underlying callback mechanisms to expand interactive use of the figure window using the Qt or other GUI backend. These work with the Jupyter notebook using the ipympl module and the %matplotlib widget cell magic.

6.1.19   Keyboard Events

The Matplotlib GUI figure window is capable of listening and responding to keyboard events (i.e., typed keystrokes) when the figure window has focus. The `mpl_connect` and `mpl_disconnect` functions attach listeners to the events in the GUI window and trigger the corresponding callback when changes are detected. Note the use of the `sys.stdout.flush()` to clear the standard output.

```python
import sys
fig, ax = plt.subplots()
# disconnect default handlers
fig.canvas.mpl_disconnect(fig.canvas.manager.key_press_handler_id)
ax.set_title('Keystroke events',fontsize=18)
def on_key_press(event):
    print (event.key)
    sys.stdout.flush()

# connect function to canvas and listen
fig.canvas.mpl_connect('key_press_event', on_key_press)
plt.show()
```

Now, you should be able to type in keys in the figure window and see some output in the terminal. We can make the callback interact with the artists in the figure window by referencing them, as in the following example:

```python
fig, ax = plt.subplots()
x = np.arange(10)
line, = ax.plot(x, x*x,'-o') # get handle of Line2D object
ax.set_title('More Keystroke events',fontsize=18)
def on_key_press(event):
    # If event.key is one of shorthand color notations, set line
    ↪  color
    if event.key in 'rgb':
        line.set_color(event.key)
        fig.canvas.draw() # force redraw

# disconnect default figure handlers
fig.canvas.mpl_disconnect(fig.canvas.manager.key_press_handler_id)
# connect function to canvas and listen
fig.canvas.mpl_connect('key_press_event', on_key_press)
plt.show()
```

You can also increase the marker size using this same technique by tacking on another callback as in the following example:

```python
def on_key_press2(event):
    # If the key is one of the shorthand color notations,
    set the line color
    if event.key in '123':
        val = int(event.key)*10
        line.set_markersize(val)
        fig.canvas.draw() # force redraw

fig.canvas.mpl_connect('key_press_event', on_key_press2)
```

You can also use modifier keys like alt, ctrl, shift.

```
import re
def on_key_press3(event):
    'alt+1,alt+2, changes '
    if re.match('alt\+?',event.key):
        key,=re.match('alt\+(.?)',event.key).groups(0)
        val = int(key)/5.
        line.set_mew(val)
        fig.canvas.draw() # force redraw

fig.canvas.mpl_connect('key_press_event', on_key_press3)
plt.show()
```

Now, you can use the given numbers, letters, and modifiers in your keystrokes to alter the embedded line. Note that there is also a on_key_press_release event if you want to hook into multiple letters or have a noisy keyboard.

6.1.20 Mouse Events

You can also tap into mouse movement and clicks in the figure window using a similar mechanism. These events can respond with figure/data coordinates and an integer ID of the button (i.e., left/middle/right) pressed.

```
fig, ax = plt.subplots()
# disconnect default handlers
fig.canvas.mpl_disconnect(fig.canvas.manager.key_press_handler_id)

def on_button_press(event):
    button_dict = {1:'left',2:'middle',3:'right'}
    print ("clicked %s button" % button_dict[ event.button ])
    print ("figure coordinates:", event.x, event.y)
    print ("data coordinates:", event.xdata, event.ydata)
    sys.stdout.flush()

fig.canvas.mpl_connect('button_press_event', on_button_press)
plt.show()
```

Here is another version that puts points at every click point.

```
fig, ax = plt.subplots()
ax.axis([0,1,0,1])
# disconnect default handlers
fig.canvas.mpl_disconnect(fig.canvas.manager.key_press_handler_id)

o=[]
def on_button_press(event):
    button_dict = {1:'left',2:'middle',3:'right'}
    ax.plot(event.xdata,event.ydata,'o')
    o.append((event.xdata,event.ydata))
    sys.stdout.flush()
    fig.canvas.draw()
```

```
fig.canvas.mpl_connect('button_press_event', on_button_press)
plt.show()
```

In addition to clicks, you can also position the mouse over an artist on the canvas and click on it to access that artist. This is known as the `pick` event.

```
fig, ax = plt.subplots()
ax.axis([-1,1,-1,1])
ax.set_aspect(1)
for i in range(5):
   x,y= np.random.rand(2).T
   circle = Circle((x, y),radius=0.1 , picker=True)
   ax.add_patch(circle)

def on_pick(event):
    artist = event.artist
    artist.set_fc(np.random.random(3))
    fig.canvas.draw()

fig.canvas.mpl_connect('pick_event', on_pick)
plt.show()
```

By clicking on the circles in the figure, you can randomly change their corresponding colors. Note that you have to set the `picker=True` keyword argument when the artist is instantiated. You can also do `help(plt.connect)` to use other events.

6.2   Seaborn

Seaborn facilitates specialized plots that make it quick and easy to visually assess statistical aspects of data. Because Seaborn is built on top of Matplotlib, it can target any output that Matplotlib supports. Seaborn is extraordinarily well-written with fantastic documentation and easy-to-read source code. We can import Seaborn in the usual way, noting that we also want to load the traditional `plt` interface to Matplotlib.

```
>>> import pandas as pd
>>> import matplotlib.pyplot as plt
>>> import seaborn as sns
```

Seaborn comes with many interesting datasets,

```
>>> tips = sns.load_dataset('tips')
>>> tips.head()
   total_bill  tip     sex smoker  day    time  size
0       16.99 1.01  Female     No  Sun  Dinner     2
1       10.34 1.66    Male     No  Sun  Dinner     3
2       21.01 3.50    Male     No  Sun  Dinner     3
3       23.68 3.31    Male     No  Sun  Dinner     2
4       24.59 3.61  Female     No  Sun  Dinner     4
```

Fig. 6.31 Scatterplot for
`tips` dataset

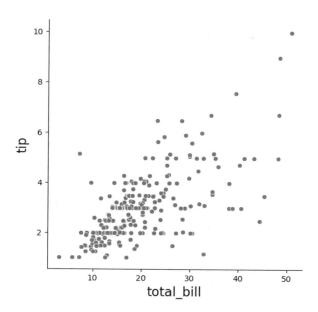

We can easily investigate relations between pairs of variables using `sns.`
`relplot()` in Fig. 6.31,

```
>>> sns.relplot(x='total_bill',y='tip',data=tips)
<seaborn.axisgrid.FacetGrid object at 0x7f9370b261c0>
```

Although we could easily re-create this plot in plain Matplotlib using `scatter`,
the Seaborn practice is to pass in the Pandas dataframe `tips` and reference the
columns to be plotted with the `x` and `y` keyword arguments. Then, other dataframe
columns can be assigned to other graphical *roles* in the rendering that permit layers
of graphical concepts onto a single figure. Further, the background and frame
are stylistically different than the Matplotlib defaults. Importantly, the resulting
object returned is a `seaborn.axisgrid.FacetGrid` instance so if we want to
retrieve the native Matplotlib elements, we have to extract them from `FacetGrid`
as the attributes, `fig` and `ax`, for example.

By assigning the `hue` role to the `smoker` column of the dataframe, the resulting
plot is the same as before except the individual circles are colored by the `smoker`
categorical column of the dataframe, as in Fig. 6.32,

```
>>> sns.relplot(x='total_bill',hue='smoker',y='tip',data=tips)
<seaborn.axisgrid.FacetGrid object at 0x7f9370c3f400>
```

Instead of changing the color for each smoker category, we can assign the `smoker`
column to the `style` role, which changes the shape of the marker in Fig. 6.33,

```
>>> sns.relplot(x='total_bill',y='tip',
...             style='smoker', # different marker shapes
...             data=tips)
<seaborn.axisgrid.FacetGrid object at 0x7f9372fe1700>
```

Fig. 6.32 Individual data
points colored using the
categorical column `smoker`

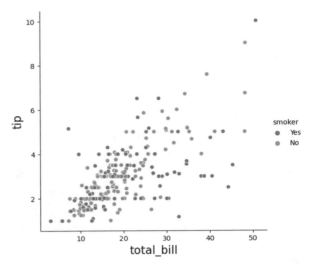

Fig. 6.33 Instead of color,
different markers can be used
with the categorical `smoker`
column

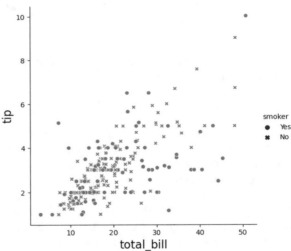

You can also specify any valid Matplotlib marker type for each of the two categorical
values of the `smoker` column in Fig. 6.34 by specifying the `markers` keyword
argument,

```
>>> sns.relplot(x='total_bill',y='tip',
...             style='smoker',
...             markers=['s','^'],data=tips)
<seaborn.axisgrid.FacetGrid object at 0x7f9376c6c310>
```

The `size` column of the input dataframe can be used to scale each of the markers
in Fig. 6.35,

Fig. 6.34 Custom markers can be specified for individual data points using the categorical `smoker` column

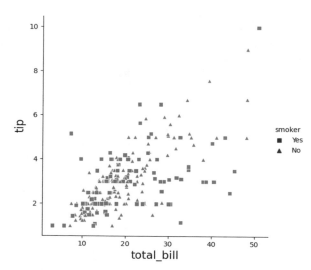

Fig. 6.35 The `size` categorical column in the `tips` dataset can scale the individual points

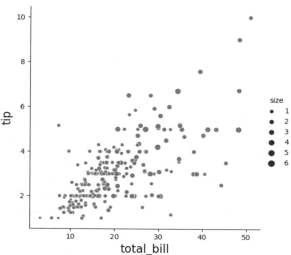

```
>>> sns.relplot(x='total_bill',y='tip',
...             size='size', # scale markers
...             data=tips)
<seaborn.axisgrid.FacetGrid object at 0x7f93713bd310>
```

This is discrete because of the unique values of that column in the dataframe,

```
>>> tips['size'].unique()
array([2, 3, 4, 1, 6, 5])
```

These sizes can be scaled between bounds specified by the `sizes` keyword argument, such as `sizes=(5,10)`. Note that supplying a continuous numerical

Fig. 6.36 Options such as
transparency (i.e., alpha
value) that are not used by
Seaborn can be passed down
to the Matplotlib renderer as
keyword arguments

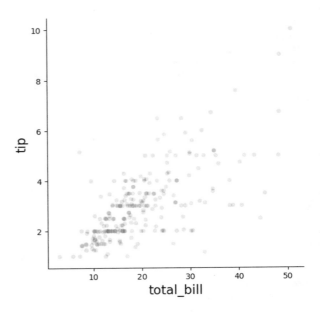

column like `total_bill` for the `size` keyword argument will bin that column
automatically, instead of allowing for a continuous range of marker sizes. The
transparency value of each point is specified by the `alpha` keyword argument
which is passed down to Matplotlib to render. This means Seaborn does not use
it to filter the options that flow down to Matplotlib `scatter` as in Fig. 6.36.

```
>>> sns.relplot(x='total_bill',y='tip',alpha=0.3,data=tips)
<seaborn.axisgrid.FacetGrid object at 0x7f937127b640>
```

Further, because Seaborn processes the `size` keyword argument it is not possible
to push the `scatter` size keyword argument through to the final Matplotlib
rendering. Thus, the trade-off is that Seaborn makes specific editorial changes on
the appearance of the resulting graphic that may be hard to tweak outside of Seaborn
semantics. The `scatter` plot is the default `kind` for `relplot()` but a line plot
can be used instead with the `kind='line'` keyword argument.

6.2.1 Automatic Aggregation

If a dataset has multiple `x` values with different `y-values`, then Seaborn will
aggregate these values to produce a containing envelope for the resulting line.

```
>>> fmri = sns.load_dataset('fmri')
>>> fmri.head()
   subject  timepoint event    region  signal
0      s13         18  stim  parietal   -0.02
```

```
1        s5         14   stim   parietal    -0.08
2        s12        18   stim   parietal    -0.08
3        s11        18   stim   parietal    -0.05
4        s10        18   stim   parietal    -0.04
```

There are fifty-six `signal` values for each of the `timepoint` values,

```
>>> fmri['timepoint'].value_counts()
18     56
8      56
1      56
2      56
3      56
4      56
5      56
6      56
7      56
9      56
17     56
10     56
11     56
12     56
13     56
14     56
15     56
16     56
0      56
Name: timepoint, dtype: int64
```

Here are some statistics on the `timepoint=0` group,

```
>>> fmri.query('timepoint==0')['signal'].describe()
count     56.00
mean      -0.02
std        0.03
min       -0.06
25%       -0.04
50%       -0.02
75%        0.00
max        0.07
Name: signal, dtype: float64
```

Because of this multiplicity, Seaborn will plot the mean of each of the points as shown with the `marker='o'` with confidence intervals above and below specifying the 95% confidence interval of the mean estimate as in Fig. 6.37,

```
>>> sns.relplot(x='timepoint', y='signal', kind='line',data=fmri,
↪    marker='o')
<seaborn.axisgrid.FacetGrid object at 0x7f93709c5df0>
```

The downside is that the estimate is computed using bootstrapping, which can be slow for large datasets and can be turned off using the `ci=None` keyword argument or replaced with `ci='sd'`, which will instead use the easier-to-compute standard deviation. The `estimator=None` keyword argument will make this stop altogether. As with `kind='scatter'`, additional dataframe columns can be

Fig. 6.37 Data containing multiple *x*-values for differing *y*-values can be automatically aggregated with bootstrapped estimate of confidence intervals

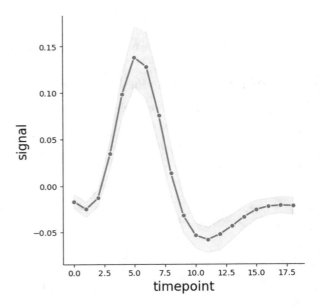

recruited to distinguish the resulting graphic. The following uses `hue='event'` to draw distinct line plots for each of the categories of `fmri.event`,

```
>>> fmri.event.unique()
array(['stim', 'cue'], dtype=object)
```

Now, we have distinct line plots for each `fmri.event` categorical as in Fig. 6.38,

```
>>> sns.relplot(x='timepoint', y='signal',
...             kind='line', data=fmri,
...             hue='event', marker='o')
<seaborn.axisgrid.FacetGrid object at 0x7f93709c5df0>
```

If the `hue` column is numerical instead of categorical, then the colors will be scaled over a continuous interval. Consider the following dataset and corresponding Fig. 6.39:

```
>>> dots = sns.load_dataset('dots').query('align == "dots"')
>>> dots.head()
  align choice  time  coherence  firing_rate
0  dots     T1   -80       0.00        33.19
1  dots     T1   -80       3.20        31.69
2  dots     T1   -80       6.40        34.28
3  dots     T1   -80      12.80        32.63
4  dots     T1   -80      25.60        35.06
>>> sns.relplot(x='time', y='firing_rate',
...             hue='coherence', style='choice',
...             kind='line', data=dots)
<seaborn.axisgrid.FacetGrid object at 0x7f9370883cd0>
```

There are two different line styles because there are two different unique values in the `choice` column. The lines are colored according to the numerical `coherence`

Fig. 6.38 The categorical `event` column can color each dataset differently

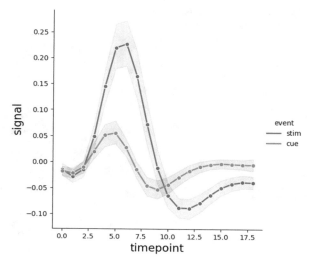

Fig. 6.39 When columns are numerical instead of categorical, the colors can be scaled along a continuous range

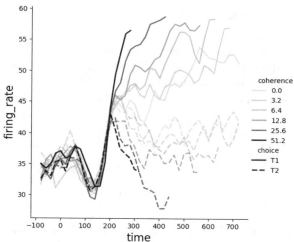

column. Seaborn is particularly strong with color palettes. The following generates a colormap over `n_colors`:

```
>>> palette = sns.color_palette('viridis', n_colors=6)
```

Then, re-drawing the same plot in Fig. 6.40 makes the individually colored lines somewhat more distinct, especially with a slightly thicker line width,

```
>>> sns.relplot(x='time', y='firing_rate',palette=palette,
...             hue='coherence', style='choice',
...             kind='line', data=dots, linewidth=2)
<seaborn.axisgrid.FacetGrid object at 0x7f937085a130>
```

Fig. 6.40 Same as Fig. 6.39
except with different colors

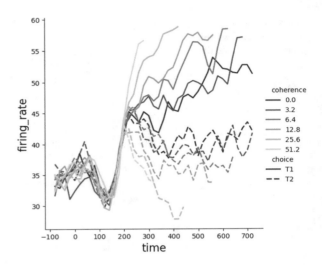

6.2.2   Multiple Plots

Because `relplot` returns a `FacetGrid` object, it is straightforward to create
multiple subplots be specifying the `row` and `col` keyword arguments. The following Fig. 6.41 stacks the two subplots side-by-side in two columns because there are
two unique values for the `time` column in the dataframe.

```
>>> tips.time.unique()
['Dinner', 'Lunch']
Categories (2, object): ['Dinner', 'Lunch']

>>> sns.relplot(x='total_bill', y='tip', hue='smoker',
...             col='time', # columns for time
...             data=tips)
<seaborn.axisgrid.FacetGrid object at 0x7f9370704ee0>
```

Multiple gridded plots can be generated along multiple rows and columns. The
`col_wrap` keyword argument keeps the following Fig. 6.42 from filling very wide
graph and instead puts each of the subplots into separate rows,

```
>>> sns.relplot(x='timepoint', y='signal', hue='event',
...             style='event',
...             col='subject', col_wrap=5,
...             height=3, aspect=.75, linewidth=2.5,
...             kind='line',
...             data=fmri.query('region == "frontal"'))
<seaborn.axisgrid.FacetGrid object at 0x7f9370895be0>
```

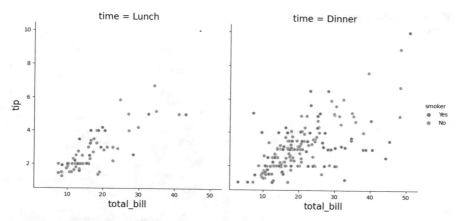

Fig. 6.41 Matplotlib subplots are called *facets* in Seaborn

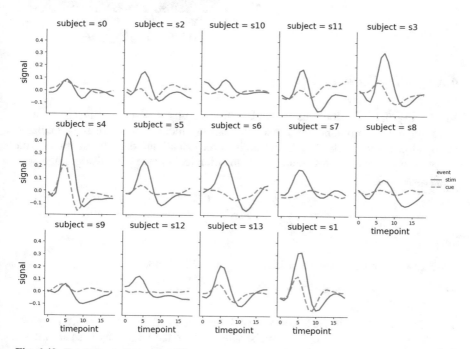

Fig. 6.42 Facets support complex tilings

Fig. 6.43 Seaborn supports
histograms

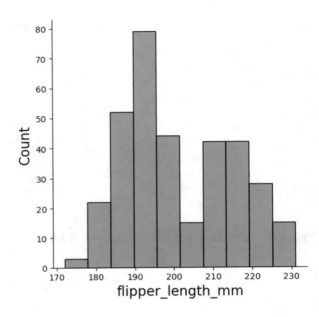

6.2.3   *Distribution Plots*

Visualizing data distributions is fundamental to any statistical analysis as well
as to diagnosing problems with machine learning models. Seaborn is particularly
strong at providing well-thought-out options for visualizing data distributions. The
simplest univariate data distribution visualization technique is the histogram,

```
>>> penguins = sns.load_dataset("penguins")
>>> penguins.head()
  species     island  bill_length_mm  bill_depth_mm  flipper_length_mm  body_mass_g     sex
0  Adelie  Torgersen           39.10          18.70             181.00      3750.00    Male
1  Adelie  Torgersen           39.50          17.40             186.00      3800.00  Female
2  Adelie  Torgersen           40.30          18.00             195.00      3250.00  Female
3  Adelie  Torgersen             nan            nan                nan         nan     NaN
4  Adelie  Torgersen           36.70          19.30             193.00      3450.00  Female
```

Seaborn's `displot` draws a quick histogram in Fig. 6.43,

```
>>> sns.displot(penguins, x="flipper_length_mm")
<seaborn.axisgrid.FacetGrid object at 0x7f9370951b20>
```

The width of the bins in the histogram can be selected with the `binwidth` keyword
argument and the number of bins can be selected with the `bins` keyword argument.
For categorical data with a few distinct values, bins can be selected as a sequence of
distinct values, like `bins=[1,3,5,8]`. Alternatively, Seaborn can automatically
handle this with the `discrete=True` keyword argument.

Other columns can be used to create overlapping semi-transparent histograms as
in the following Fig. 6.44:

```
>>> sns.displot(penguins, x="flipper_length_mm", hue="species")
<seaborn.axisgrid.FacetGrid object at 0x7f9363c04f70>
```

Fig. 6.44 Multiple histograms can be overlaid using colors and transparency

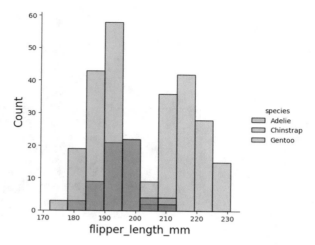

Fig. 6.45 Same as Fig. 6.44 except with no distracting vertical lines

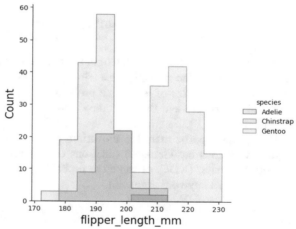

Seaborn intelligently selects the bins to match for each respective histogram, which is hard to do with Matplotlib's default `hist()` function.

Sometimes the vertical lines of the histogram can become distracting. These can be removed with the `element='step'` keyword argument in Fig. 6.45,

```
>>> sns.displot(penguins, x='flipper_length_mm',
...                        hue='species',
...                        element='step')
<seaborn.axisgrid.FacetGrid object at 0x7f9363b01f70>
```

Instead of layering using transparency, histograms can be stacked on top of each other (see Fig. 6.46) but this can be hard to reconcile with many histograms and a few *dominant* histograms that may overwhelm or obscure the others.

```
>>> sns.displot(penguins, x="flipper_length_mm",
...                        hue="species",
```

Fig. 6.46 Multiple
histograms can be stacked

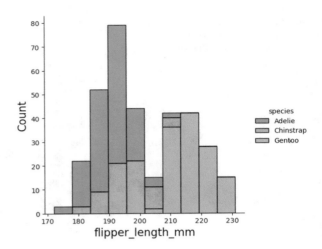

```
...                              multiple="stack")
<seaborn.axisgrid.FacetGrid object at 0x7f9372f6fbb0>
```

Alternatively, if there are few histograms, the bars can be narrowed enough to fit side-by-side using the multiple='dodge' keyword argument. For histograms to be used as estimates of legitimate probability density functions, they have to be scaled appropriately and this is done via the stat='probability' keyword argument.

Histograms approximate the univariate probability density function with rectangular functions but Gaussian functions can be used to create smoother Kernel Density Estimates (KDE) of same the probability density function in Fig. 6.47 by using the kind keyword argument.

```
>>> sns.displot(penguins, x="flipper_length_mm", kind="kde")
<seaborn.axisgrid.FacetGrid object at 0x7f936398f3d0>
```

The smoothness of the resulting plot is determined by the KDE bandwidth parameter which is available as the bw_adjust keyword argument. Keep in mind that even though smoothness may be visually pleasing, it may be mis-representing discontinuities or suppressing features in the data that may be important.

Beyond univariate probability density functions, Seaborn provides powerful visualization tools for bivariate probability density functions (see Fig. 6.48).

```
>>> sns.displot(penguins, x="bill_length_mm", y="bill_depth_mm")
<seaborn.axisgrid.FacetGrid object at 0x7f9363635730>
```

Figure 6.48 shows the two-dimensional grid which tallies the number of data points in each with corresponding color scale for these counts. Using the kind='kde' keyword argument works with bivariate distributions also. If there is not a lot of overlap between the bivariate distributions, then these can be plotted in different colors using the hue keyword argument (see Fig. 6.49). Different bin-widths for

Fig. 6.47 Seaborn also
supports Kernel Density
Estimates (KDEs)

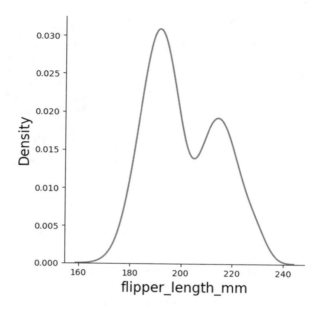

Fig. 6.48 Bivariate
histogram

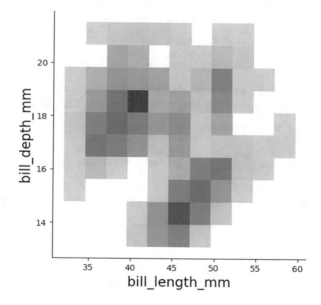

Fig. 6.49 Multiple bivariate histograms can be drawn together with different colors

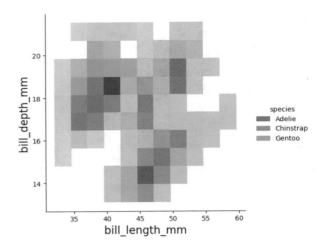

each coordinate dimension can be selected using the `binwidth` keyword argument but with a tuple defining the bin width for each coordinate dimension. The `cbar` keyword argument draws the corresponding color scale but when combined with the `hue` keyword argument, it will draw multiple color scales (one for each category encoded with `hue`), which makes the resulting Fig. 6.49 very crowded.

```
>>> sns.displot(penguins, x='bill_length_mm',
...                 y='bill_depth_mm', hue='species')
<seaborn.axisgrid.FacetGrid object at 0x7f93632f2040>
```

Plotting the marginal distributions of a bivariate distribution is very helpful and Seaborn makes that easy with the `jointplot` as in Fig. 6.50.

```
>>> sns.jointplot(data=penguins,
...                 x="bill_length_mm",
...                 y="bill_depth_mm")
<seaborn.axisgrid.JointGrid object at 0x7f9362f25670>
```

The `jointplot` returns a `JointGrid` object which allows for drawing the marginals separately using the `plot_marginals()` method for Fig. 6.51,

```
>>> g = sns.JointGrid(data=penguins,
...                 x="bill_length_mm",
...                 y="bill_depth_mm")
>>> g.plot_joint(sns.histplot)
<seaborn.axisgrid.JointGrid object at 0x7f93632e8f10>
>>> g.plot_marginals(sns.kdeplot,fill=True)
<seaborn.axisgrid.JointGrid object at 0x7f93632e8f10>
```

Note that the marginals are now KDE plots instead of box plots. Unused keyword arguments in `plot_marginals()` are passed down to the function argument (`fill=True` for `sns.kdeplot` in this case). Beyond `jointplot`, Seaborn provides `pairplot` which will produce a `jointplot` for all pairs of columns in the input dataframe.

Fig. 6.50 Marginal distributions corresponding to bivariate distributions can be drawn along the horizontal/vertical axes

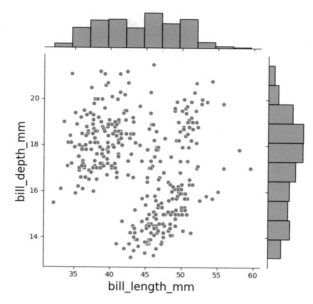

Fig. 6.51 Customized marginal distributions can also be drawn

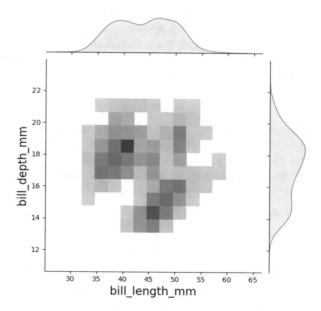

Fig. 6.52 Discrete
distributions along
categorical variables

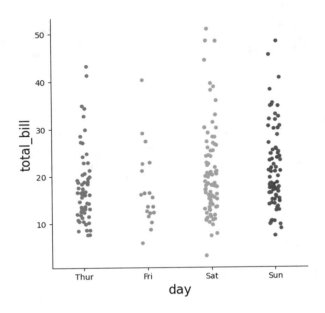

Seaborn provides many options for visualizing categorical plots. Whereas we have been mainly using categories to provide distinct colors or overlay plots, Seaborn provides specialized visualizations specifically for categorical data. For example, the following jittered plot in Fig. 6.52 shows the scatter of the `total_bill` based on the day of the week.

```
>>> sns.catplot(x="day", y="total_bill", data=tips)
<seaborn.axisgrid.FacetGrid object at 0x7f93623c2610>
```

Importantly, the individual points are randomly scattered along the horizontal direction to avoid overlaying and obscuring the points. This can be turned off using the `jitter=False` keyword argument.

An interesting method instead of using jitter to avoid obscuring data is to use a `swarmplot` by using the `kind='swarm'` keyword argument in Fig. 6.53,

```
>>> sns.catplot(data=tips,
...             x="day",
...             y="total_bill",
...             hue="sex",
...             kind="swarm")
<seaborn.axisgrid.FacetGrid object at 0x7f93623cb220>
```

The extent of the extended Christmas tree-like arms in Fig. 6.53 replaces the random positioning due to jitter. One fascinating aspect of this plot is the clustering effect it produces which automatically draws your attention to certain features that would be hard to ascertain beforehand. For example, Fig. 6.53 seems to show that the total bill for males is greater than for females on Saturday, and, in particular, is around `total_bill=20`. Using Seaborn's `displot`, we can verify that quickly with the following code for Fig. 6.54:

Fig. 6.53 Same as Fig. 6.52
but using a *swarm* plot

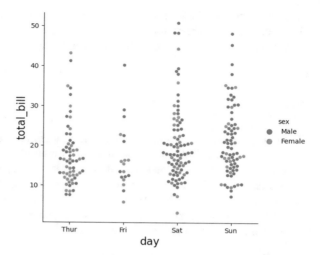

Fig. 6.54 Histogram shows
details of swarm plot

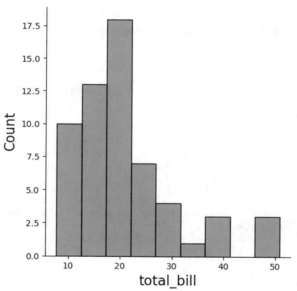

```
>>> sns.displot(tips.query('day=="Sat" and sex=="Male"'),
...             x='total_bill')
<seaborn.axisgrid.FacetGrid object at 0x7f9362021a90>
```

Colors, lines, and general presentation of Seaborn graphics can be controlled
with `sns.set_theme()` on a global level or on a figure-specific level by using a
context manager like `with sns.axes_style("white")`, for example. Finer-grain
control of fonts and other details are available with the `set()` function. Sequences
of colors for discrete data or clear distinguishable colors for continuous data are

strongly supported in Seaborn, as we have seen previously with `color_palette` . The human perceptual system responds strongly to colors and many of the common pitfalls have been reconciled in Seaborn's approach to managing colors for data.

6.3 Bokeh

The combination of web-based visualization technologies and Python opens up a wide-range of possibilities for web-based interactive visualization using modern Javascript frameworks. This makes it possible to deploy Python data visualizations with users interacting via modern web browsers such as Google Chrome or Mozilla Firefox (not Internet Explorer!). Bokeh is an open-source Python module that makes it much easier to develop and deploy such interactive web-based visualizations because it provides primitives that sidestep writing low-level Javascript.

6.3.1 Using Bokeh Primitives

Here is a quick example that generates Fig. 6.55:

```
from bokeh.plotting import figure, output_file, show
# prepare some data
x = range(10)
y = [i**2 for i in x]
# create a new plot with a title and axis labels
p = figure(title="plot the square", # title of figure
           x_axis_label='x',
           y_axis_label='y',
           width= 400,  # figure width
           height = 300) # figure height
# add a line renderer with legend and line thickness
p.line(x, y,
       legend=r"x^2", # text that appears in legend
       line_width=2)  # width of line

# show the results
show(p)
```

Bokeh follows the same philosophy as Matplotlib in terms of layering the graphical elements on the canvas. Because Bokeh can render to multiple endpoints (including Jupyter notebooks), we choose the endpoint as an output html file with the `output_file` function. Once that is settled, the main step is to create a figure canvas with the `figure` function and supply parameters for the figure like the `x_axis_label` and `width`. Then, we add the line with the `line` method on the figure object include line parameters like `line_width`. Finally, the `show` function actually writes the HTML to the file specified in `output_file`. Thus, the key

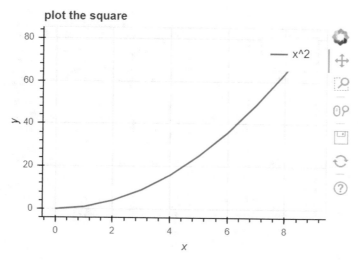

Fig. 6.55 Basic Bokeh plot

benefit of Bokeh is that it allows staging the interactive visualization in Python while rendering the necessary Javascript into the HTML output file.

On the right margin of the figure, we have the default tools for the figure. The pan tool at the top drags the line around with the figure frame. The zoom tool is just below that. These tools are embedded in the HTML as Javascript functions that are run by the browser. This means that Bokeh created both the static HTML *and* the embedded Javascript. The data for the plot is also likewise embedded in the so-rendered HTML document. Like Matplotlib, Bokeh provides low-level control of every element that is rendered on the canvas. Simultaneously, this means you can create powerful and compelling graphics using the primitives in the toolkit at the cost of programming every element in the graphic. Like Matplotlib, Bokeh also supports a gallery[1] of detailed example graphics you can use as starting points for your own visualizations.

You can specify the individual properties of a graphical element (i.e., glyph) when it is added to the canvas. The next line adds red circles of a specified radius to mark each of the data points (Fig. 6.56):

```
p.circle(x,y,radius=0.2,fill_color='red')
```

You can also add properties after the fact by saving the individual object as in the following:

```
c = p.circle(x,y)
c.glyph.radius= 0.2
c.glyph.fill_color = 'red'
```

---

[1] https://docs.bokeh.org/en/latest/docs/gallery.html.

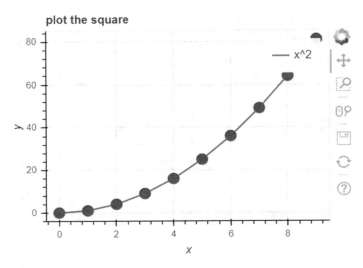

Fig. 6.56 Plot with customized markers for each data point

To add multiple glyphs to your figure, use the figure object methods (e.g., `p.circle()`, `p.line()`, `p.square()`). The main documentation [1] has a comprehensive list of available primitives. It is just a matter of learning the vocabulary of available glyphs and assigning their properties

6.3.2 Bokeh Layouts

Analogous to the `subplots` function in Matplotlib, Bokeh has the `row` and `column` functions that arrange individual figures on the same canvas. The next example uses the `row` function to put two figures side-by-side (Fig. 6.57).

```
from bokeh.plotting import figure, show
from bokeh.layouts import row, column
from bokeh.io import output_file
output_file('bokeh_row_plot.html')
x = range(10)
y = [i**2 for i in x]
f1 = figure(width=200,height=200)
f1.line(x,y,line_width=3)
f2 = figure(width=200,height=200)
f2.line(x,x,line_width=4,line_color='red')
show(row(f1,f2))
```

In this example, we created the individual figure objects and then used the `show()` method on the `row()` function to create the side-by-side layout of the two `f1` and `f2` figures. This can be combined with the `column` function to create grid layouts as in the following (Fig. 6.58):

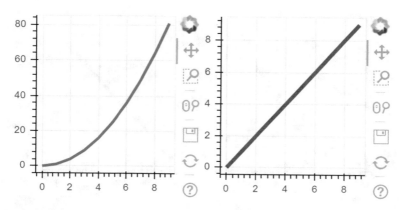

Fig. 6.57 Bokeh row plot

```
from bokeh.plotting import figure, show
from bokeh.layouts import row, column
from bokeh.io import output_file

output_file('bokeh_row_column_plot.html')

x = range(10)
y = [i**2 for i in x]
f1 = figure(width=200,height=200)
f1.line(x,y,line_width=3)
f2 = figure(width=200,height=200)
f2.line(x,x,line_width=4,line_color='red')
f3 = figure(width=200,height=200)
f3.line(x,x,line_width=4,line_color='black')
f4 = figure(width=200,height=200)
f4.line(x,x,line_width=4,line_color='green')
show(column(row(f1,f2),row(f3,f4)))
```

Note that because the individual figures are treated separately, they have their own
set of tools. This can be alleviated using `bokeh.layouts.gridplot` or the
more general `bokeh.layouts.layout` function.

6.3.3 Bokeh Widgets

Interactive widgets allow the user to interact and explore the Bokeh visualization.
The main site has the inventory of available widgets. These are primarily based on
the Bootstrap JavaScript library. All widgets have the same two-step implementation
pattern. The first step is to create the widget as it will be laid out on the canvas. The
second step is to create the callback that will be triggered when the user interacts
with the widget.

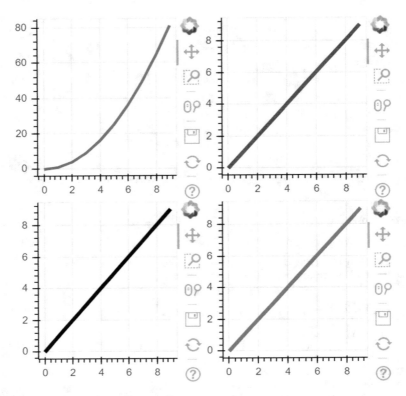

Fig. 6.58 Bokeh row-column plot

Here is where things get tricky. Given that the callback specifies some kind of action based on the widget, where is that action going to take place? If the output is a HTML file that is rendered by the browser, then that action must be handled via Javascript and run in the browser. The complications start when that interaction is driven by or dependent upon objects in the Python workspace. Remember that after the HTML file has been created, there is no more Python. If you want callbacks that utilize a Python process, you have to use bokeh serve to host that application (Fig. 6.59).

Let us start with the callbacks that are handled by Javascript in the static HTML output, as in the following:

```
from bokeh.io import output_file, show
from bokeh import events
from bokeh.models.widgets import Button
from bokeh.models.callbacks import  CustomJS
output_file("button.html")
cb = CustomJS(args=dict(),code='''
    alert("ouch!");
    ''')
button = Button(label='Hit me!') # create button object
```

Fig. 6.59 JavaScript callback
for button widget

Fig. 6.60 JavaScript callback
for drop-down widget

```
button.js_on_event(events.ButtonClick, cb)
show(button)
```

The key step here is the CustomJS function which takes the embedded string of valid Javascript and packages it for the static HTML file. The button widget does not have any arguments so the args variable is just an empty dictionary. The next important part is the js_on_event function which specifies the event (i.e., ButtonClick) and the callback that is assigned to handle that event. Now, when the output HTML file is created and you render the page in the browser and then click on the button named *Hit me!* you will get a popup from the browser with the text *ouch* in it.

Following this structure, we can try something more involved, as in the following (Fig. 6.60):

```
from bokeh.io import output_file, show
from bokeh.models.callbacks import  CustomJS
from bokeh.layouts import widgetbox
from bokeh.models.widgets import Dropdown
output_file("bokeh_dropdown.html")
cb = CustomJS(args=dict(),code='alert(cb_obj.value)')
menu = [("Banana", "item_1"),
        ("Apple", "item_2"),
        ("Mango", "item_3")]
dropdown = Dropdown(label="Dropdown button", menu=menu,callback=cb)
show(widgetbox(dropdown))
```

In this example, we are using the DropDown widget and populating it with the items in the menu. The first element of the tuple is the string that will appear in the dropdown and the second element is the Javascript variable that is associated with it. The callback is the same alert as before except now with the cb_obj variable. This is the callback object that is automatically passed into the Javascript function when it is activated. In this case, we specified the callback in the callback keyword argument instead of relying on the event types as before. Now, after the pages rendered in the browser and you click and pulldown on an element in the menu, you should see the popup window.

The following example combines the pulldown and the line graph we created earlier. The only new element here is passing the line object into the CustomJS function via the args keyword argument. Once inside the Javascript code, we can change the embedded line_color of the line glyph based on the value selected in the pulldown via the cb_obj.value variable. This allows us to have a widget control the color property of the line. Using the same approach, we can alter other

Fig. 6.61 Javascript callback
updates line color

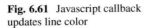

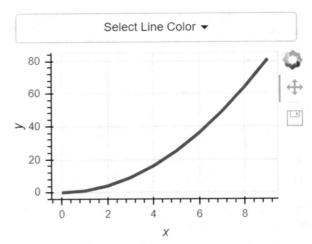

properties of other objects as long as we pass them in via the `args` keyword
argument to the `CustomJS` function (Fig. 6.61).

```
from bokeh.plotting import figure, output_file, show
from bokeh.layouts import column
from bokeh.models.widgets import Dropdown
from bokeh.models.callbacks import  CustomJS
output_file("bokeh_line_dropdown.html")
# make some data
x = range(10)
y = [i**2 for i in x]
# create the figure as usual
p = figure(width = 300, height=200, # figure width and height
           tools="save,pan", # only these tools
           x_axis_label='x', # label x and y axes
           y_axis_label='y')
# add the line,
line = p.line(x,y,               # x,y data
              line_color='red', # in red
              line_width=3)     # line thickness
# elements for pulldown
menu = [("Red", "red"),("Green", "green"),("Blue", "blue")]
# pass the line object and change line_color based on pulldown
↪   choice
cb = CustomJS(args=dict(line=line),
              code='''
                   var line = line.glyph;
                   var f = cb_obj.value;
                   line.line_color = f;
                   ''')
# assign callback to Dropdown
dropdown = Dropdown(label="Select Line Color",
↪   menu=menu,callback=cb)
# use column to put Dropdown above figure
show(column(dropdown,p))
```

Fig. 6.62 Javascript callback
updates graph sine wave
frequency

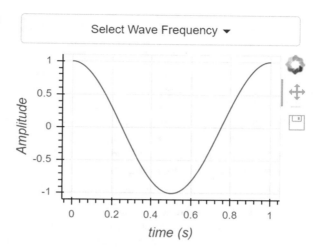

The last example shows how to manipulate line properties using browser-executed callback. Bokeh can go deeper by altering the data itself and then letting the embedded figure react to those changes. The next example is very similar to the last example except here we use the `ColumnDataSource` object to ferry data into the browser from Python. Then, in the embedded Javascript code, we unpack the data and then change the data array based on the action in the pulldown. The crucial part is to trigger the data update and draw using the `source.change.emit()` function (Fig. 6.62).

```
from bokeh.plotting import figure, output_file, show
from bokeh.layouts import row, column
from bokeh.models import ColumnDataSource
from bokeh.models.widgets import Dropdown
from bokeh.models.callbacks import  CustomJS
import numpy as np
output_file("bokeh_ColumnDataSource.html")
# make some data
t = np.linspace(0,1,150)
y = np.cos(2*np.pi*t)
# create the ColumnDataSource and pack the data in it
source = ColumnDataSource(data=dict(t=t, y=y))
# create the figure as usual
p = figure(width = 300, height=200, tools="save,pan",
           x_axis_label='time (s)',y_axis_label='Amplitude')
# add the line, but now using the ColumnDataSource
line = p.line('t','y',source=source,line_color='red')
menu = [("1 Hz", "1"), ("5 Hz", "5"), ("10 Hz", "10")]
cb = CustomJS(args=dict(source=source),
              code='''
                    var data = source.data;
                    var f = cb_obj.value;
                    var pi = Math.PI;
                    t = data['t'];
                    y = data['y'];
```

```
                    for (i = 0; i < t.length; i++) {
                        y[i] = Math.cos(2*pi*t[i]*f)
                    }
                    source.change.emit();
                    ''')
dropdown = Dropdown(label="Select Wave Frequency",
↪   menu=menu,callback=cb)
show(column(dropdown,p))
```

The advantage of this approach is that it creates a self-contained HTML file that no longer needs Python. On the other hand, it could be that the callback requires active Python computation that is not possible for in-the-browser JavaScript. In this case, we can use the bokeh server as in

```
Terminal> bokeh serve bokeh_ColumnDataSource_server.py
```

which will run a small server and open a web page pointing to the local page. The web page contains basically the same content as the last example, although you may notice that the page is less responsive than the last example. This is because the interaction has to round-trip back to the server process and then back to the browser instead of updating in the browser itself. The contents of bokeh_ColumnDataSource_server are shown below

```
from bokeh.plotting import figure, show, curdoc
from bokeh.layouts import column
from bokeh.models import ColumnDataSource, Select
from bokeh.models.widgets import Dropdown
import numpy as np
# make some data
t = np.linspace(0,1,150)
y = np.cos(2*np.pi*t)
# create the ColumnDataSource and pack the data in it
source = ColumnDataSource(data=dict(t=t,y=y))

# create the figure as usual
p = figure(width = 300, height=200, tools="save,pan",
            x_axis_label='time (s)',y_axis_label='Amplitude')
# add the line, but now using the ColumnDataSource
line = p.line('t','y',source=source,line_color='red')
menu = [("1 Hz", "1"),
        ("5 Hz", "5"),
        ("10 Hz", "10")]
def cb(attr,old,new):
    f=float(freq_select.value)
    d = dict(x=source.data['t'],y=np.cos(2*np.pi*t*f))
    source.data.update(d)

freq_select = Select(value='1 Hz', title='Frequency (Hz)',
options=['1','5','10'])
freq_select.on_change('value',cb)
curdoc().add_root(column(freq_select,p))
```

Note the Select element takes the place of the DropDown menu. The freq_select.on_change('value',cb) is how the pulldown selection is communicated to the Python server process. Importantly, the options keyword argument takes a sequence of strings ['1','5','10'] even though we have to convert the strings back to float in the callback.

```
freq_select = Select(value='1 Hz',
                      title='Frequency (Hz)',
                      options=['1','5','10'])
```

The final key step is to include `curdoc().add_root(column(freq_select,p))` which creates an oriented `column` that is attached to the document and that will respond to the callback. Note there are many configuration options for deploying a `bokeh` server, including using Tornado as a backend and running multiple threads (see main documentation for details). Bokeh also has many integration options with Jupyter notebook.

Bokeh is under active development and the overall design and architecture of Bokeh is well thought-out. The combination of JavaScript and Python visualizations vended through web browsers is a topic of incredible growth in the open-source community. Bokeh represents just one (and arguably the best) of many implementations of this joint strategy. Stay tuned for further development and features from the Bokeh team, but bear in mind that in contrast to Matplotlib, Bokeh is less mature.

6.4 Altair

Another project in the space of web-based scientific visualizations is `Altair` which is a *declarative* visualization module. Declarative means the data elements are assigned a visual role that Altair implements; in contrast to Matplotlib wherein all the details of the construction have to be specified (i.e., imperative visualization). Altair implements the Vega-Lite JSON graphical grammar for visualization. This means that the low-level rendered graphics are actually implemented in JavaScript via `Vega`. The best way to work with Altair is via Jupyter notebooks as creating stand-alone visualizations takes a number of additional steps, details of which are on the main documentation site. When using Jupyter Notebook, you may have to enable the Altair renderer,

```
import altair as alt
```

The main Altair object is the `Chart` object.

```
>>> import altair as alt
>>> from altair import Chart
>>> import vega_datasets
>>> cars = vega_datasets.data('cars')
>>> chart = Chart(cars)
```

Importantly, `data` input to `Chart` should be a Pandas dataframe, `altair.Data` object, or a URL referencing a CSV or JSON file in so-called *tidy* format, which basically means that the columns are variables and the rows of the dataframe are observations of those variables. To create the Altair visualization, you have to decide on a mark (i.e., glyph) for the data values. Doing `chart.mark_point()` will just draw a circle in the Jupyter notebook. To draw a plot, you have to specify

Fig. 6.63 One-dimensional
Altair chart

Fig. 6.64 Two-dimensional
Altair chart

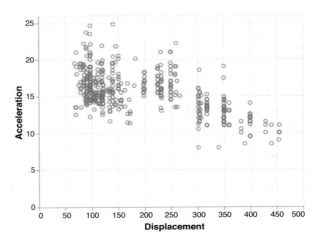

channels for the other elements of the dataframe as in the following corresponding
to Fig. 6.63:

```
>>> chart.mark_point().encode(x='Displacement')
alt.Chart(...)
```

which creates a one-dimensional scatterplot using the `Displacement` column of
the input dataframe. Conceptually, `encode` means create a mapping between the
data and its visual representation. Using this pattern, to create a two-dimensional
X-Y plot, we need to specify an additional channel for the y-data as in the following
(see Fig. 6.64):

```
>>> chart.mark_point().encode(x = 'Displacement',
...                           y = 'Acceleration')
alt.Chart(...)
```

This creates a two-dimensional plot of acceleration versus displacement. Note that
the `chart` object has access to the column names in the input dataframe so that
we can access them by their string names as keyword arguments to the `encode`
method. Because each row has a corresponding categorical name, we can use that as
the color dimension in the so-created X-Y plot using the `color` keyword argument
as in the following Fig. 6.65:

```
>>> chart.mark_point().encode(x='Displacement',
...                           y='Acceleration',
...                           color='Origin')
alt.Chart(...)
```

The key to using Altair is to realize that it is a thin layer that leverages the
Vega-Lite JavaScript visualization library. Thus, you can build any kind of Vega-

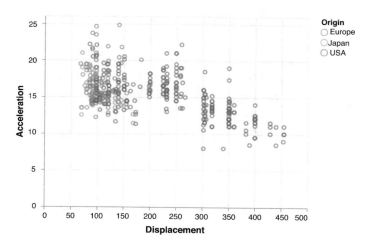

Fig. 6.65 The color of each maker is determined by the `Origin` column in the input dataframe

lite visualization in Altair and connect it to Pandas dataframes or other Python constructs without having to write the JavaScript yourself. If you want to change the individual colors or line widths of the glyphs, then you have to use the Vega editor and then pull the Vega-lite specification back into Altair, which is not difficult, but involves multiple steps. The point is that this kind of customization is not what Altair is good for. As a declarative visualization module, the idea is to leave those details to Altair and concentrate on mapping the data to visual elements.

6.4.1 Detailing Altair

Controlling the individual elements of the visual presentation in Altair can be accomplished using the family of `configure_` top level functions as in Fig. 6.66.

```
>>> (chart.configure_axis(titleFontSize=20,
...                       titleFont='Consolas')
...       .mark_point()
...       .encode(x='Displacement',
...               y='Acceleration',
...               color='Origin')
... )
alt.Chart(...)
```

Note that the `titleFontSize` for the titles on each of the axes has increased and the corresponding font family has been updated. Instead of taking the default column names as the title labels, we can use the `alt.X()` and `alt.Y()` objects to customize these as in the following Fig. 6.67:

```
>>> (chart.configure_axis(titleFontSize=20,
...                       titleFont='Consolas')
```

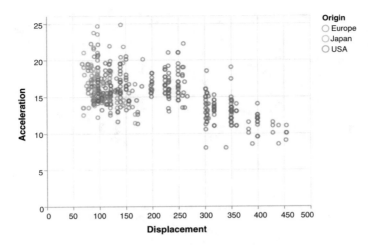

Fig. 6.66 The title font family and size can be set with `configure_axis`

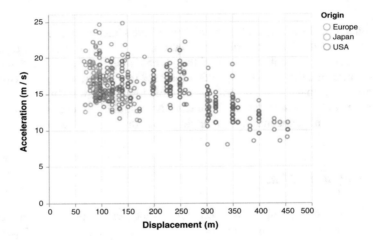

Fig. 6.67 With `alt.X` and `alt.Y`, the labels on each axis can be changed

```
...             .mark_point()
...             .encode(x=alt.X('Displacement',
...                             title='Displacement (m)'),
...                     y=alt.Y('Acceleration',
...                             title='Acceleration (m/s)'),
...                     color='Origin')
... )
alt.Chart(...)
```

Note that each of the labels has units. If we wanted to control the vertical axis, we could have specified `configure_axisLeft` and only that axis would have been affected by these changes. There are also `configure_axisTop` and

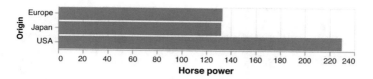

Fig. 6.68 Barcharts are generated by `mark_bar` on the named x and y columns in the dataframe

`configure_axisRight`, `configure_axisX`, `configure_axisY` also. The labeling of the tick marks on the axis is controlled via the `labelFontSize` and other related parameters.

Barcharts are generated following the same pattern except with `mark_bar`, as in the following Fig. 6.68:

```
>>> chart.mark_bar().encode(y='Origin',x='Horsepower')
alt.Chart(...)
```

Likewise, area plots are generated using `mark_area` and so forth. You can stick `interactive()` at the end of the call to make the plot zoomable with the mouse-wheel. Charts can be automatically saved into PNG and SVG formats but these require additional browser automation tools or `altair_saver`. Charts can also be saved into HTML files with the requisite JavaScript embeddings.

6.4.2 Aggregations and Transformations

Altair can perform certain aggregations on data elements from a dataframe. This saves you from having to add the aggregations into another separate dataframe that would have to be passed as input (see Fig. 6.69).

```
>>> cars = vega_datasets.data('cars')
>>> (alt.Chart(cars).mark_point(size=200,
...                             filled=True)
...             .encode(x='Year:T',
...                     y='mean(Horsepower)')
... )
alt.Chart(...)
```

The text `mean` in the string means that the average of the Horsepower will be used for the y-value (see Fig. 6.69). The suffix `:T` means that the `Year` column in the dataframe should be treated as a *timestamp*. The arguments to the `mark_point` function control the size and fill properties of the point markers.

In the prior Fig. 6.69, we computed the mean of the `Horsepower` but included all vehicle origins. If we wanted to include only US-made vehicles, then we can use the `transform_filter` method, as in the following Fig. 6.70:

```
>>> (alt.Chart(cars).mark_point(size=200,filled=True)
...             .encode(x='Year:T',
```

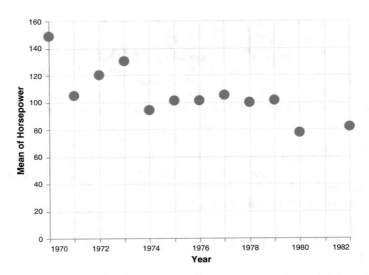

Fig. 6.69 Aggregations like taking the `mean` are available via Altair

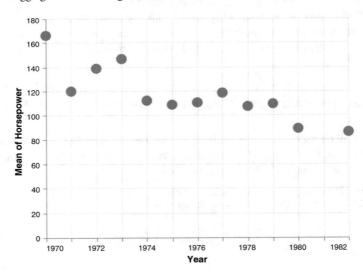

Fig. 6.70 Filters can be used in the aggregations with `transform_filter`

```
...                        y='mean(Horsepower)',
...                        )
...           .transform_filter('datum.Origin=="USA"')
... )
alt.Chart(...)
```

The filter transformation ensures that only the US vehicles are concluded in the mean calculation. The word `datum` is how Vega-lite refers to its data elements. In

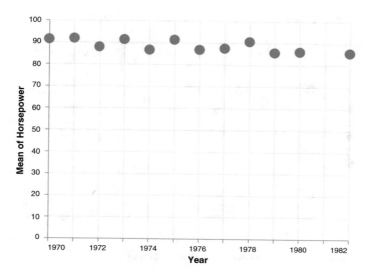

Fig. 6.71 The alt.FieldRangePredicate filters a range of values

addition to expressions, Altair provides predicate objects that can perform advanced filtering operations. For example, using the FieldRangePredicate, we can select out a range of values from a continuous variable as in the following Fig. 6.71:

```
>>> (alt.Chart(cars).mark_point(size=200,filled=True)
...     .encode(x='Year:T',
...             y='mean(Horsepower)',
...             )
...     .transform_filter(alt.FieldRangePredicate('Horsepower',
...                                               [75,100]))
... )
alt.Chart(...)
```

This means that only Horsepower values between 75 and 100 will be considered in the calculation and the resulting Altair chart (Fig. 6.72).

```
>>> (alt.Chart(cars).mark_point(size=200,
...                 filled=True)
...     .encode(x='Year:T',
...             y=alt.Y('mean(Horsepower)',
...             scale=alt.Scale(domain=[60,110])),
...             )
...     .transform_filter(alt.FieldRangePredicate('Horsepower',
...                                               [75,100]))
...     .properties(width=300,height=200)
... )
alt.Chart(...)
```

Note that the scale limits of the resulting figure have been adjusted using alt.Scale with the domain keyword. Predicates can be combined using logical operations like LogicalNotPredicate and others.

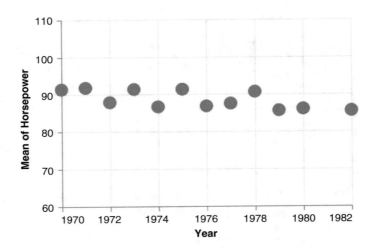

The `transform_calculate` method allows basic expressions to be applied to data elements. For example, to calculate the square of the horsepower, we can do the following Fig. 6.73:

```
>>> from altair import datum
>>> h1=(alt.Chart(cars).mark_point(size=200,
...                                filled=True)
...                    .encode(x='Year:T',
...                            y='sqh:Q')
...                    .transform_calculate(sqh =
↪    datum.Horsepower**2)
... )
```

Note that the type of the resulting calculation has to be specified using the `:Q` (for *quantitative*) in reference to the resultant.

We can also use the `transform_aggregate` method to use the resulting `sqh` variable from the previous `transform_calculate` method to compute the mean over the squares as shown below while grouping over the `Year` as in Fig. 6.74,

```
>>> h2=(alt.Chart(cars).mark_point(size=200,
...                                filled=True,
...                                color='red')
...                    .encode(x='Year:T',
...                            y='msq:Q')
...                    .transform_calculate(sqh =
↪    datum.Horsepower**2)
...                    .transform_aggregate(msq='mean(sqh)',
...                                         groupby=['Year'])
... )
```

These two charts can be overlaid using the + operator (Fig. 6.75),

```
>>> h1+h2
alt.LayerChart(...)
```

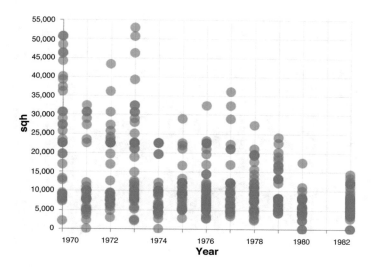

Fig. 6.73 Embedded calculations are implemented with `transform_calculate`

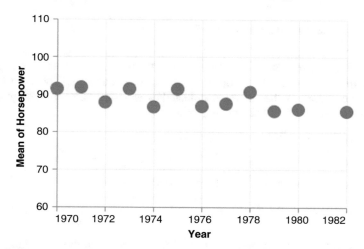

Fig. 6.74 Transform and aggregations compute intermediate items in Altair visualizations

There are other functions in the `transform_*` family including `transform_lookup`, `transform_window`, and `transform_bin`. The main documentation site has details.

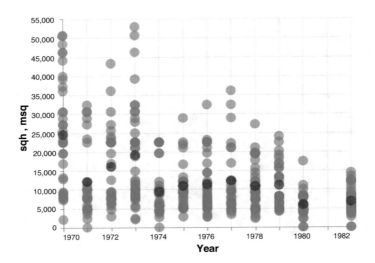

Fig. 6.75 Charts can be overlaid with the plus operator

6.4.3 Interactive Altair

Altair has interactive features that are inherited from Vega-lite. These are expressed as tools that can be easily assigned to Altair visualizations. The core component of interactions is the *selection* object. For example, the following creates a selection_interval() object (see Fig. 6.76),

```
>>> brush = alt.selection_interval()
```

Now, we can attach the brush to an Altair chart (Fig. 6.76).

```
>>> chart.mark_point().encode(y='Displacement',
...                           x='Horsepower')\
...       .properties(selection=brush)
alt.Chart(...)
```

From within the web browser, you can drag the mouse in the chart and you will see a rectangular selection box appear, but nothing will change because we have not attached the brush selector to an *action* on the chart. To change the color of the so-selected objects on the chart, we can use the alt.condition method in the encode method (Fig. 6.77),

```
>>> (chart.mark_point().encode(y='Displacement',
...                            x='Horsepower',
...                            color=alt.condition(brush,
...                                                'Origin:N',
...                                     alt.value('lightgray')))
...       .properties(selection=brush)
... )
alt.Chart(...)
```

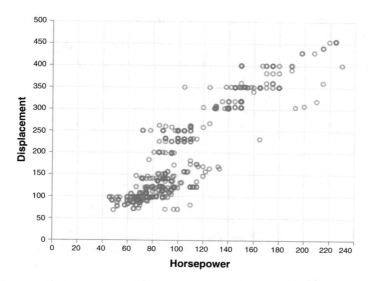

Fig. 6.76 Interactive selection can be enabled with `alt.selection_interval`

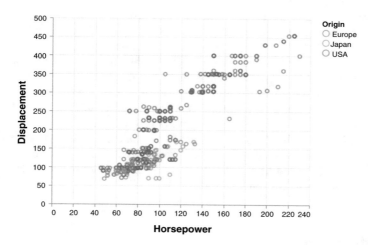

Fig. 6.77 Selecting elements in the chart triggers `alt.Condition`, which changes how the elements are rendered

The condition means that if the items in the brush selection is `True`, then the points are colored according to their `Origin` values and otherwise set to `lightgray` using the `alt.value`, which sets the value to use for the encoding.

Selectors can be shared across charts, as in the following (Fig. 6.78):

```
>>> chart = (alt.Chart(cars)
...             .mark_point()
...             .encode(x='Horsepower',
...                     color=alt.condition(brush,
...                                          'Origin:O',
...                                          alt.value('lightgray')))
...             .properties(selection=brush)
...          )
>>> chart.encode(y='Displacement') & chart.encode(y='Acceleration')
alt.VConcatChart(...)
```

A number of subtle things have happened here. First, note that the `chart` variable only has the x coordinate defined and that the `cars` data is embedded. The vertical ampersand `&` symbol stacks the two vertically. It is only on the last line that the y coordinate for each chart is selected. The selector is used for both charts so that selecting using the mouse in either one of them will cause the selection (via the `alt.condition`) to highlight on both charts. This is a complicated interaction that took very few lines of code!

The other selection objects like `alt.selection_multi` and `alt.selection_single` allow for individual item selection using mouse click or mouse hover actions (e.g., `alt.selection_single(on='mouseover')`). Selections for `alt.selection_multi` require mouse-clicking while holding down the shift key.

Altair is a novel and smart take on the fast-changing visualization landscape. By piggy-backing on Vega-lite, the module has ensured that it can keep up with that important body of work. Altair brings new visualizations that are not part of the standard Matplotlib or Bokeh vocabulary into Python. Still, Altair is far less mature than either Matplotlib or Bokeh. Vega-lite itself is a complicated package with its own JavaScript dependencies. When something breaks, it is hard to fix because the problem may be too deep and spread out in the JavaScript stack for a Python programmer to reach (or even identify!).

6.5 Holoviews

Holoviews provide a means for annotating data to facilitate downstream visualization using Bokeh, Plotly, or Matplotlib. The key concept is to provide semantics to the data elements that enable construction of data visualizations. This a declarative approach similar to Altair but it is agnostic with respect to the downstream visualization construction whereas `altair` has a fixed downstream target of Vega-Lite.

```
>>> import numpy as np
>>> import pandas as pd
```

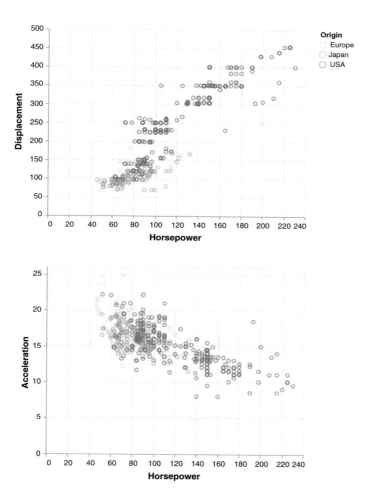

Fig. 6.78 Interactive selectors can be shared across Altair charts

```
>>> import holoviews as hv
>>> from holoviews import opts
>>> hv.extension('bokeh')
```

The hv.extension part declares that Bokeh should be used as the downstream visualization constructor. Let us create some data,

```
>>> xs = np.linspace(0,1,10)
>>> ys = 1+xs**2
>>> df = pd.DataFrame(dict(x=xs, y=ys))
```

To visualize this dataframe with Holoviews we have to decide on what we want to render and how we want to render it. One way to do that is using a Holoviews Curve (see Fig. 6.79).

Fig. 6.79 Plotting data using
Holoviews `Curve`

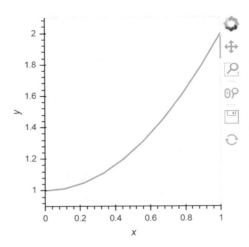

```
>>> c=hv.Curve(df,'x','y')
>>> c
:Curve     [x]     (y)
```

The Holoviews `Curve` object declares that the data is from a continuous function
that maps x into y columns in the dataframe. To render this plot, we use the built-in
Jupyter display,

```
>>> c
:Curve     [x]     (y)
```

Note that the `Bokeh` widgets are already included in the plot. The `Curve` object
still retains the embedded dataframe which can be manipulated via the `data`
attribute. Note that we could have supplied Numpy arrays, Python lists, or a Python
dictionary instead of the Pandas dataframe.

```
>>> c.data.head()
      x      y
0  0.00  1.00
1  0.11  1.01
2  0.22  1.05
3  0.33  1.11
4  0.44  1.20
```

We can choose Holoviews `Scatter` instead of `Curve` if we want to *not* fill line
segments between the data points (see Fig. 6.80),

```
>>> s = hv.Scatter(df,'x','y')
>>> s
:Scatter    [x]     (y)
```

We can combine these two plots side-by-side in Fig. 6.81 using the overloaded +
operator,

Fig. 6.80 Plotting use
Holoviews `Scatter`

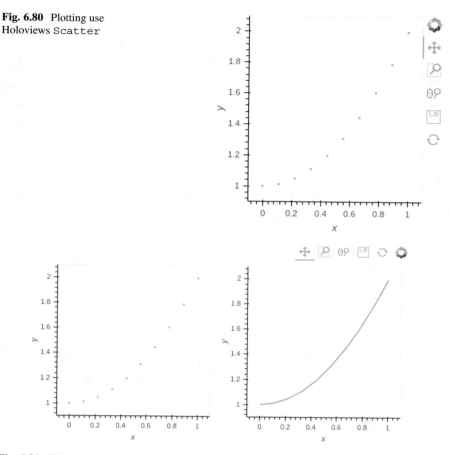

Fig. 6.81 Side-by-side layouts via the addition operator

```
>>> s+c # notice how the Bokeh tools are shared between plots and
↪    affect both
:Layout
   .Scatter.I :Scatter   [x]   (y)
   .Curve.I   :Curve   [x]   (y)
```

These can be overlaid using the overloaded ∗ operator (Fig. 6.82),

```
>>> s*c # overlay with multiplication operator
:Overlay
   .Scatter.I :Scatter   [x]   (y)
   .Curve.I   :Curve   [x]   (y)
```

Options like `width` and `height` on the figure can be sent directly as shown in
Fig. 6.83

```
>>> s.opts(color='r',size=10)
:Scatter   [x]   (y)
```

Fig. 6.82 Overlaid plots with multiplication operator

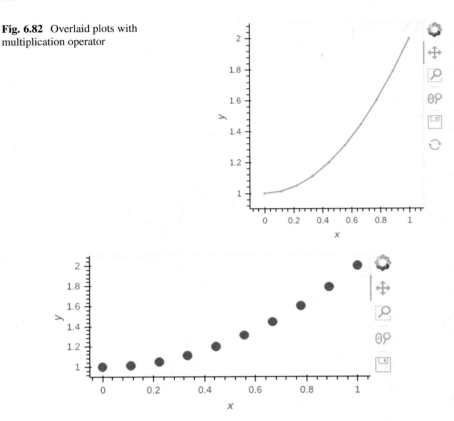

Fig. 6.83 Visualization dimensions can be set using width and height in the options method

```
>>> s.options(width=400,height=200)
:Scatter   [x]   (y)
```

You can also use the %opts magic within the Jupyter notebook to manage these options. While this is a very well organized approach, it only works within the Jupyter notebook. To reset the title as shown in Fig. 6.84, we use the relabel method.

```
>>> k = s.redim.range(y=(0,3)).relabel('Ploxxxt Title') # put a title
>>> k.options(color='g',size=10,width=400,
...               height=200,yrotation=88)
:Scatter   [x]   (y)
```

There are many other options for plotting as shown in the reference gallery. Figure 6.85 is an example using Bars. Note the use of the to method to create the Bars object from the Scatter object. The xrotation alters the orientation of the x-axis ticklabels.

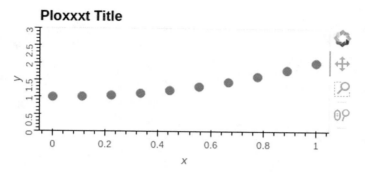

Fig. 6.84 Visualization titles are set using `relabel` on the Holoviews object

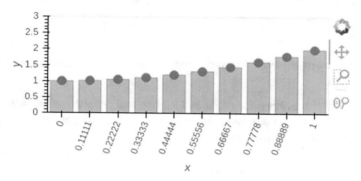

Fig. 6.85 Barplots are available in Holoviews

```
>>> options = (opts.Scatter(width=400,
...                         height=300,
...                         xrotation=70,
...                         color='g',
...                         size=10),
...            opts.Bars(width=400,
...                      height=200,
...                      color='blue',
...                      alpha=0.3)
...            )
>>> (s.to(hv.Bars)*s).redim.range(y=(0,3)).options(*options)
:Overlay
   .Bars.I     :Bars    [x]    (y)
   .Scatter.I  :Scatter    [x]    (y)
```

Data can be *sliced* from the Holoviews object where the indexing is based on the x
coordinate and None means until the end of the x data. Recall that the plus operator
combines the plots into Fig. 6.86,

```
>>> s[0:0.5] + s.select(x=(0.5,None)).options(color='b')
:Layout
   .Scatter.I   :Scatter    [x]    (y)
   .Scatter.II  :Scatter    [x]    (y)
```

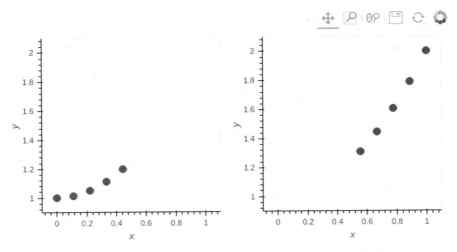

Fig. 6.86 The underlying data in Holoviews graphs can be sliced and automatically rendered

Holoviews references *key* dimensions and kdims and *value* dimensions as vdims.
These categorize which elements are to be considered independent or dependent
variables (respectively) in the plot.

```
>>> s.kdims,s.vdims
([Dimension('x')], [Dimension('y')])
```

6.5.1   Dataset

A Holoviews Dataset is *not* built-in metadata that enables visualization. Rather,
a Dataset is a way of creating a set of dimensions that will later be inherited by
the downstream visualizations that are created from the Dataset. In the following,
only the key dimensions need to be specified. The other columns are inferred to be
value dimensions (vdims), which can later be grouped using groupby.

```
>>> economic_data = pd.read_csv('macro_economic_data.csv')
>>> edata = hv.Dataset(data=economic_data,
...                          kdims=['country','year'])

>>> edata.groupby('year')
:HoloMap    [year]
   :Dataset    [country]      (growth,unem,capmob,trade)
>>> edata.groupby('country')
:HoloMap    [country]
   :Dataset    [year]      (growth,unem,capmob,trade)
```

Note that the groupby output shows the independent variable (i.e., key dimension) year or country that corresponds to any of the other value dimensions

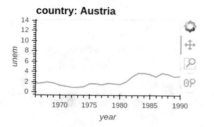 

Fig. 6.87 Holoviews can group and plot data in one step

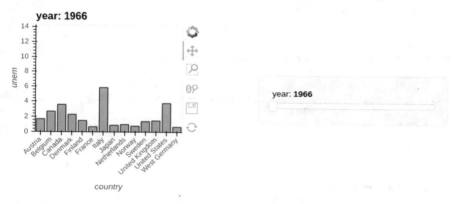

Fig. 6.88 Holoviews automatically chooses the slider widget based on the type of data in the remaining dimension

(dependent variables). In Fig. 6.87, the `to` method creates a corresponding pulldown widget for `country`.

```
>>> edata.to(hv.Curve,
...          'year',
...          'unem',
...          groupby='country').options(height=200)
:HoloMap   [country]
   :Curve  [year]     (unem)
```

We can also group by and other key dimension with a corresponding slider for `year` as shown in Fig. 6.88. Note that the type of widget was inferred from the type of the key dimension (i.e., slider for continuous `year` and pulldown for discrete `country`).

```
>>> (edata.sort('country')
...       .to(hv.Bars,'country','unem',groupby='year')
...       .options(xrotation=45,height=300))
:HoloMap   [year]
   :Bars   [country]    (unem)
```

In Fig. 6.89, we do not use any of the declared key dimensions so that they *both* go into corresponding widgets.

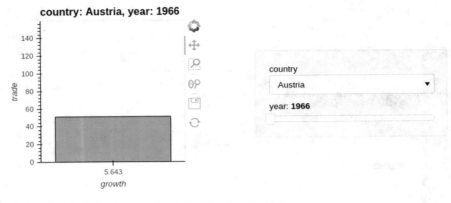

Fig. 6.89 Holoviews chooses widgets based on key dimensions

```
>>> edata.sort('country').to(hv.Bars,'growth','trade')
:HoloMap   [country,year]
   :Bars   [growth]   (trade)
```

6.5.2   Image Data

Holoviews elements can handle two-dimensional tabular or grid-based data, such as images (see Fig. 6.90).

```
>>> x = np.linspace(0, 10, 500)
>>> y = np.linspace(0, 10, 500)
>>> z = np.sin(x[:,None]*2)*y

>>> image = hv.Image(z)
>>> image
:Image   [x,y]   (z)
```

We can get a useful histogram in Fig. 6.91 of the image colors using the `hist` method

```
>>> image.hist()
:AdjointLayout
   :Image   [x,y]   (z)
   :Histogram   [z]   (z_count)
```

6.5.3   Tabular Data

```
>>> economic_data= pd.read_csv('macro_economic_data.csv')
>>> economic_data.head()
```

Fig. 6.90 Holoviews creates
heatmap images

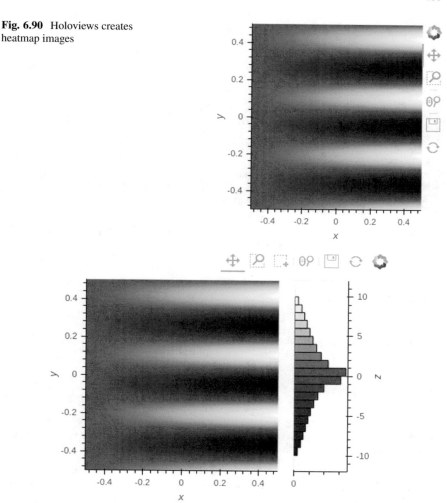

Fig. 6.91 Histograms added to images

```
           country   year   growth   unem   capmob   trade
0   United States    1966     5.11   3.80        0    9.62
1   United States    1967     2.28   3.80        0    9.98
2   United States    1968     4.70   3.60        0   10.09
3   United States    1969     2.80   3.50        0   10.44
4   United States    1970    -0.20   4.90        0   10.50
```

Grouping the elements is possible by specifying the `vdims` as a list of column
names and providing the Holoviews object with the `color_index` column name and
a `color` from the color palettes available (`hv.Palette.colormaps.keys()`).
Note that specifying the `hover` tool means the data for each element is shown when
hovering over it with the mouse (see Fig. 6.92). Be careful that all of the legend
items will not appear unless the figure is tall enough to accommodate it.

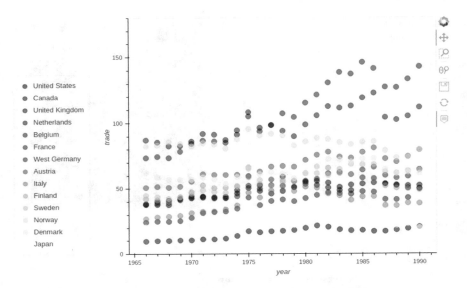

Fig. 6.92 holoviews014

```
>>> options = opts.Scatter(tools=['hover'],
...                        legend_position='left',
...                        color_index='country',
...                        width=800,height=500,
...                        alpha=0.5,
...                        color=hv.Palette('Category20'),
...                        size=10)
>>> c =
↪  hv.Scatter(economic_data,'year',['trade','country','unem'])
>>> c.redim.range(trade=(0,180)).options(options)
:Scatter   [year]    (trade,country,unem)
```

Using `Dataset` also works for two-dimensional visualizations like heatmaps.
Hovering the mouse over Fig. 6.93 shows the corresponding data for each cell.

```
>>> options = opts.HeatMap(colorbar=True,
...                        width=600,
...                        height=300,
...                        xrotation=60,
...                        tools=['hover'])

>>>
↪  edata.to(hv.HeatMap,['year','country'],'growth').options(options)
:HeatMap   [year,country]   (growth)
```

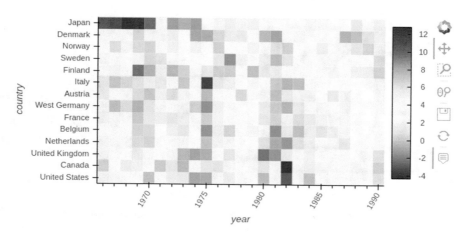

Fig. 6.93 Holoviews `Dataset` supports two-dimensional visualizations

6.5.4   Customizing Interactivity

`DynamicMap` adds interactivity to Holoviews visualizations. The `angle` key dimension is supplied by the slider widget. The named function is lazily evaluated (Fig. 6.94).

```
>>> def dynamic_rotation(angle):
...       radians = (angle / 180) * np.pi
...       return (hv.Box(0,0,4,orientation=-radians).
options(color='r',line_width=3)
...                 * hv.Ellipse(0,0,(2,4), orientation=radians)
...                 * hv.Text(0,0,'{0}°'.format(float(angle))))
...
>>> hv.DynamicMap(dynamic_rotation,
...               kdims=['angle']).redim.range(angle=(0, 360),
...                                            y=(-3,3),
...                                            x=(-3,3))
:DynamicMap     [angle]
```

Figure 6.95 is another example using corresponding slider widgets. Note that by declaring the slider widget variables ranges as floats, we obtain more resolution on the slider movement for the function.

```
>>> def sine_curve(f=1,phase=0,ampl=1):
...       xi = np.linspace(0,1,100)
...       y = np.sin(2*np.pi*f*xi+phase/180*np.pi)*ampl
...       return hv.Curve(dict(x=xi,y=y)).redim.range(y=(-5,5))
...
>>> hv.DynamicMap(sine_curve,
...               kdims=['f','phase','ampl'])\
...               .redim.range(f=(1,3.),
...                            phase=(0,360),
...                            ampl=(1,3.))
:DynamicMap     [f,phase,ampl]
```

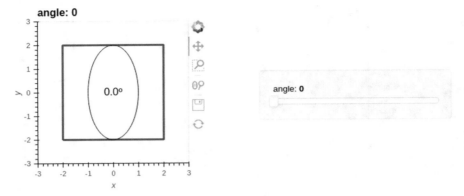

Fig. 6.94 Holoviews automatically builds widgets for specified key dimensions

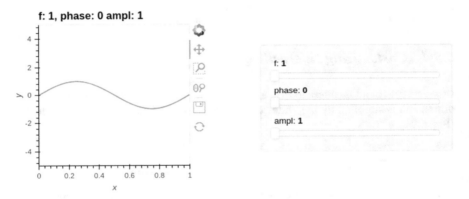

Fig. 6.95 Automatically created Holoviews widgets derive from the types of the plotted dimensions

6.5.5   Streams

Holoviews use `streams` to feed data into containers or elements instead of (for example) using the sliders for inputs. Once the stream is defined, then `hv.DynamicMap` can draw them in Fig. 6.96,

```
>>> from holoviews.streams import Stream
>>> F = Stream.define('Freq',f=3)
>>> Phase = Stream.define('phase',phase=90)
>>> Amplitude  = Stream.define('amplitude ',ampl=1)
>>> dm=hv.DynamicMap(sine_curve,streams = [F(f=1),
...                                         Phase(phase=0),
...                                         Amplitude(ampl=1)])
>>> dm
:DynamicMap    []
```

The streaming is triggered by sending the `event`,

Fig. 6.96 Holoviews
`DynamicMap` processes
streams

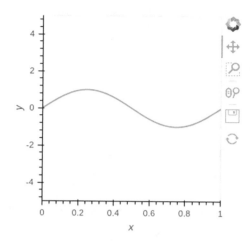

```
>>> # running causes the plot to update
>>> dm.event(f=2,ampl=2,phase=30)
```

We can also use the nonblocking `.periodic()` method,

```
>>> dm.periodic(0.1,count=10,timeout=8,param_fn=lambda
↪   i:{'f':i,'phase':i*180})
```

6.5.6   Pandas Integration with `hvplot`

Holoviews can be integrated with Pandas dataframes to accelerate common plotting scenarios by putting additional plotting options on the Pandas DataFrame object.

```
>>> import hvplot.pandas
```

Recall the following dataframe,

```
>>> df.head()
     x    y
0 0.00 1.00
1 0.11 1.01
2 0.22 1.05
3 0.33 1.11
4 0.44 1.20
```

Plots can be generated directly from the dataframe `hvplot()` method (see Fig. 6.97),

```
>>> df.hvplot() # now uses Bokeh backend instead of matplotlib
:NdOverlay   [Variable]
   :Curve    [index]    (value)
```

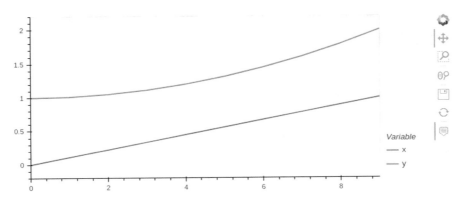

Fig. 6.97 Holoviews plots generated directly from Pandas dataframes using `hvplot`

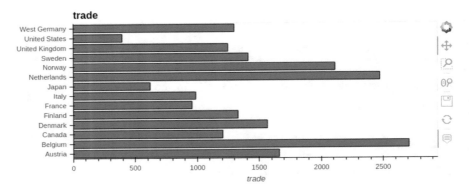

Fig. 6.98 Holoviews barcharts using `groupby`

Here is a grouping created using Pandas and drawn with Holoviews. Notice that there is no y-label on the graph. This is because the intermediate object created by the grouping is a `Series` object without a labeled column. The horizontal barchart (`barh()`) is shown in Fig. 6.98.

```
>>> economic_data.groupby('country')['trade'].sum().hvplot.barh()
:Bars   [country]   (trade)
```

Note the type of the intermediate output,

```
>>> type(economic_data.groupby('country')['trade'].sum())
<class 'pandas.core.series.Series'>
```

One way to fix the missing y-label is to cast the intervening `Series` object as a dataframe using `to_frame()` with a labeled column for the summation (see Fig. 6.99).

```
>>> (economic_data.groupby('country')['trade'].sum()
...                                            .to_frame('trade
↪   (units)')
```

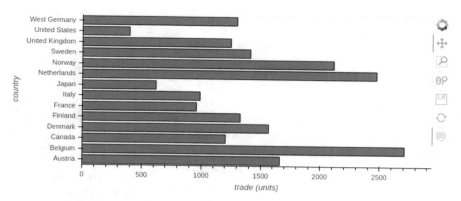

Fig. 6.99 The column names of the Pandas dataframe set labels in the visualization

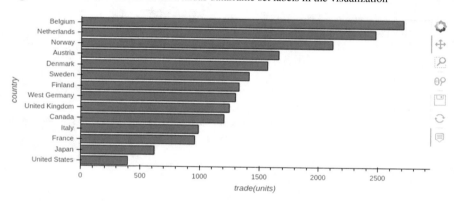

Fig. 6.100 Barcharts can be ordered using Pandas `sort_values()` method

```
 ...                                                             .hvplot.barh()
 ... )
:Bars     [country]     (trade (units))
```

We can also use Pandas to order the bars by value using `sort_values()` as shown in Fig. 6.100.

```
>>> (economic_data.groupby('country')['trade'].sum()
 ...        .sort_values()
 ...        .to_frame('trade(units)')
 ...        .hvplot.barh()
 ... )
:Bars     [country]     (trade(units))
```

Unstacking the grouping gives Fig. 6.101,

```
>>> # here is an unstacked grouping
>>> (economic_data.groupby(['year','country'])['trade']
 ...     .sum()
 ...     .unstack()
```

Fig. 6.101 holoviews025

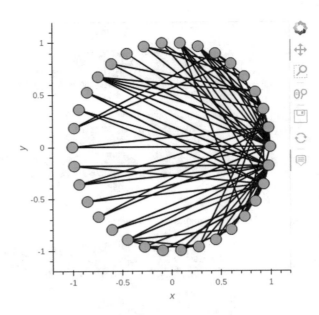

```
...       .head()
... )
country  Austria  Belgium  Canada  Denmark  ...  Sweden  United Kingdom  United States  West Germany
year                                         ...
1966     50.83    73.62    38.45   62.29     ...  44.79   37.93           9.62           37.89
1967     51.54    74.54    40.16   58.78     ...  43.73   37.83           9.98           38.81
1968     50.88    73.59    41.07   56.87     ...  42.46   37.76           10.09          39.51
1969     51.63    78.45    42.77   56.77     ...  43.52   41.93           10.44          41.40
1970     55.52    84.38    44.17   57.01     ...  46.28   42.80           10.50          43.07

[5 rows x 14 columns]
```

The barchart is missing colors for each country. Note that we have to supply the `hv.Dimension` to get the y-label scaling fixed but the `label` does not show. We also used the `rot` keyword argument to change the tick label orientation. The legend is also too tall (see Fig. 6.102).

```
>>> (economic_data.groupby(['year','country'])['trade']
...         .sum()
...         .unstack()
...         .hvplot.bar(stacked=True,rot=45)
...         .redim(value=hv.Dimension('value',label='trade',
range=(0,1000)))...   )
:Bars   [year,Variable]   (value)
```

These can be fixed in Fig. 6.103 using the cell magic with the `Variable` as the `color_index` because the dataframe does not supply a corresponding name for the values in the cells of the dataframe. The y-label is still not accessible but using `relabel`, at least we can get a title near the y-axis.

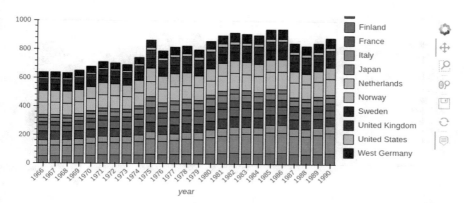

Fig. 6.102 More control for axes and labels comes from hv.Dimension

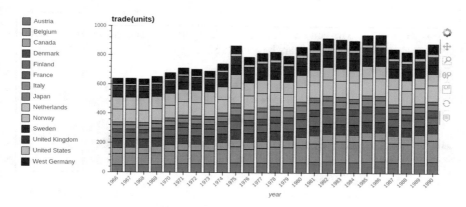

Fig. 6.103 Holoviews can draw stacked barcharts with few lines of code

```
>>> options = opts.Bars(tools=['hover'],
...                     legend_position='left',
...                     color_index='Variable',
...                     width=900,
...                     height=400)

>>> (economic_data.groupby(['year','country'])['trade']
...     .sum()
...     .unstack()
...     .hvplot.bar(stacked=True,rot=45)
...     .redim(value=hv.Dimension('value',label='trade',range=(0,1000)))
...     .relabel('trade(units)').options(options)
...    )
:Bars    [year,Variable]    (value)
```

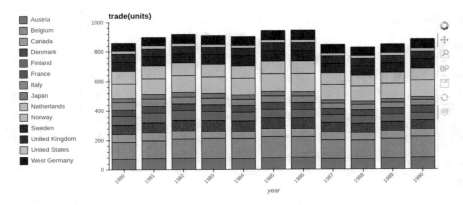

Fig. 6.104 Holoviews visualizations can be sliced like Numpy arrays

The x-range of the bars can be selected via slicing the resulting holoviews object, as shown below showing only years beyond 1980 (Fig. 6.104).

```
>>> options = opts.Bars(tools=['hover'],
...                     legend_position='left',
...                     color_index='Variable',
...                     width=900,
...                     height=400)

>>> (economic_data.groupby(['year','country'])['trade']
...      .sum()
...      .unstack()
...      .hvplot.bar(stacked=True,rot=45)
...      .redim(value=hv.Dimension('value',
...                                label='trade',
...                                range=(0,1000)))
...      .relabel('trade(units)').options(options)[1980:]
... )
:Bars    [year,Variable]    (value)
```

The main problem with the labeling with `hvplot` is that the reduce operation in the groupby does not have a name that holoviews can grab, which makes it hard to apply labels. The next block re-shapes the data into the format that `holoviews` prefers (see Fig. 6.105).

```
>>> options = opts.Bars(tools=['hover'],
...                     legend_position='left',
...                     color_index='country',
...                     width=900,
...                     stacked=True,
...                     fontsize=dict(title=18,
...                                   ylabel=16,
...                                   xlabel=16),
...                     height=400)

>>> k=(economic_data.groupby(['year','country'])['trade']
```

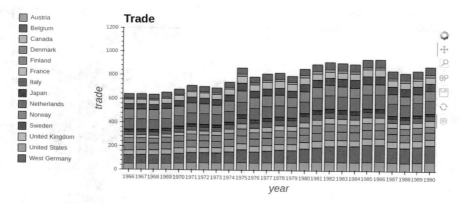

Fig. 6.105 Holoviews uses the names of Pandas columns and indices for labels

```
...        .sum().
...        to_frame().T
...        .melt(value_name='trade'))

>>> (hv.Bars(k,kdims=['year','country'])
...        .options(options)
...        .relabel('Trade').redim.range(trade=(0,1200)))
:Bars    [year,country]    (trade)
```

6.5.7 Network Graphs

Holoviews provides tools to make network graphs and integrates nicely with
`networkx`. Here are some handy defaults,

```
>>> import networkx as nx

>>> defaults = dict(width=400, height=400, padding=0.1)
>>> hv.opts.defaults(opts.EdgePaths(**defaults),
...                  opts.Graph(**defaults),
...                  opts.Nodes(**defaults))
```

We can use `networkx` to create Fig. 6.106. Note that by hovering on the circles,
you can get info about the nodes that are embedded in the graph. Note that the hover
also shows the index of the node.

```
>>> G = nx.karate_club_graph()
>>> hv.Graph.from_networkx(G,
...
↪   nx.layout.circular_layout).opts(tools=['hover'])
:Graph    [start,end]
```

Let us create the Holoviews graph separately so we can inspect it. Note that `nodes`
and `edgepaths` are how Holoviews internally represents the graph. The `nodes`

Fig. 6.106 Holoviews
supports drawing network
graphs

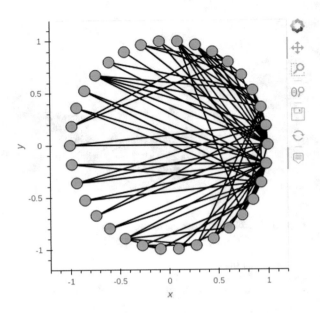

show the independent variables in the graph in brackets and the dependent variables
in parentheses. In this case it means that `club` is a node-attribute of the graph. The
`edgepaths` are the positions of the individual nodes that are determined by the
`nx.layout.circular_layout` call.

```
>>> H=hv.Graph.from_networkx(G, nx.layout.circular_layout)
>>> H.nodes
:Nodes    [x,y,index]    (club)
>>> H.edgepaths
:EdgePaths    [x,y]
```

Let us add some edge weights to the `networkx` graph and then rebuild the
corresponding Holoviews graph in Fig. 6.107,

```
>>> for i,j,d in G.edges(data=True):
...     d['weight'] = np.random.randn()**2+1
...
>>> H=hv.Graph.from_networkx(G, nx.layout.circular_layout)
>>> H
:Graph    [start,end]    (weight)
```

Note that hovering over the edges now displays the edge weights, but you can no
longer inspect the nodes themselves, as in the previous rendering. Also note that the
edge is highlighted upon hovering.

```
>>> H.opts(inspection_policy='edges')
:Graph    [start,end]    (weight)
```

You can at least color the edges based upon their weight values and choose the
colormap.

Fig. 6.107 Edge weights
added to Holoviews network
graphs

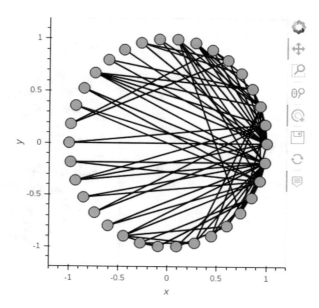

```
>>> H.opts(inspection_policy='nodes',
...          edge_color_index='weight',
...          edge_cmap='hot')
:Graph   [start,end]   (weight)
```

It is inelegant, but you can overlay two graphs using the multiplication operator with
different inspection policies to get hover info on the nodes *and* edges (Figs. 6.108
and 6.109).

```
>>> H.opts(inspection_policy='edges', clone=True) * H
:Overlay
   .Graph.I   :Graph   [start,end]   (weight)
   .Graph.II  :Graph   [start,end]   (weight)
```

You can further change the thickness of the edges using hv.dim.

```
>>> H.opts(edge_line_width=hv.dim('weight'))
:Graph   [start,end]   (weight)
```

You can use the Set1 colormap to color the nodes based upon the *club* values in
the graph.

```
>>> H.opts(node_color=hv.dim('club'),cmap='Set1')
:Graph   [start,end]   (weight)
```

We can inspect a minimum spanning tree for this graph using what we have learned
so far about Holoviews graphs in Fig. 6.110.

Fig. 6.108 Holoviews
inspection policies control
what the mouse hover shows

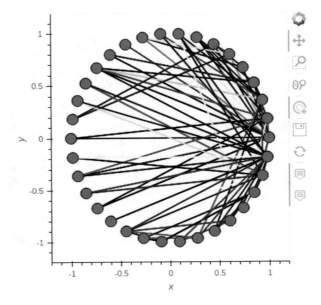

Fig. 6.109 Holoviews
network graphs can have
different edge thicknesses

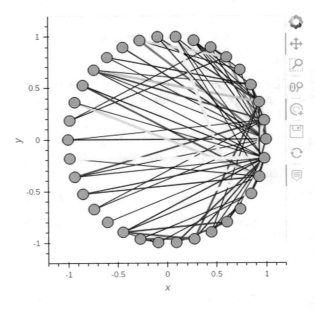

Fig. 6.110 Holoviews network graphs can have colored nodes

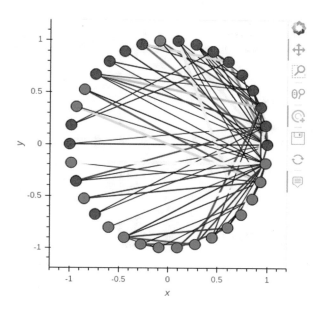

```
>>> t = nx.minimum_spanning_tree(G)
>>> T=hv.Graph.from_networkx(t, nx.layout.kamada_kawai_layout)
>>> T.opts(node_color=hv.dim('club'),cmap='Set1',
...        edge_line_width=hv.dim('weight')*3,
...        inspection_policy='edges',
...        edge_color_index='weight',edge_cmap='cool')
:Graph   [start,end]   (weight)
```

Saving Holoviz Objects Holoviews provides a hv.save() method for saving the rendered graphics into an HTML file. As long as there are no server-dependent callbacks, the so-generated HTML is a static file that is easy to distribute (Fig. 6.111).

6.5.8 Holoviz Panel for Dashboards

Panel makes dashboards with dynamically updated elements.

```
>>> import panel as pn
>>> pn.extension()
```

Similar to ipywidgets, Panel has an interact for attaching a Python callback to a widget. The following example shows how to report a value based on a slider (see Fig. 6.112):

```
>>> def show_value(x):
...     return x
...
```

Fig. 6.111 Holoviews
network graph shows a
minimum spanning tree

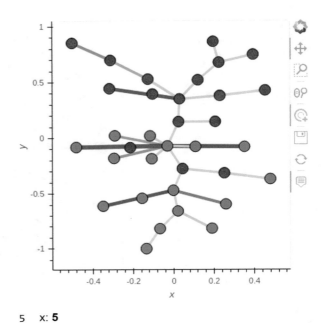

Fig. 6.112 Panel   5   x: **5**
`interact` connects
callbacks to widgets

```
>>> app=pn.interact(show_value, x=(0, 10))
>>> app
Column
    [0] Column
        [0] IntSlider(end=10, name='x', value=5, value_throttled=5)
    [1] Row
        [0] Str(int, name='interactive07377')
```

The object returned by `pn.interact` is index-able and can be re-oriented. Note
that the text appears on the left instead of at the bottom (as shown in Fig. 6.113).

```
>>> print(app)
Column
    [0] Column
        [0] IntSlider(end=10, name='x', value=5, value_throttled=5)
    [1] Row
        [0] Str(int, name='interactive07377')
>>> pn.Row(app[1], app[0]) # text and widget oriented row-wise
↪    instead of default column-wise
Row
    [0] Row
        [0] Str(int, name='interactive07377')
    [1] Column
        [0] IntSlider(end=10, name='x', value=5, value_throttled=5)
```

Panel Component Types There are three main types of components in Panel:

- Pane: A Pane wraps a view of an external object (text, image, plot, etc.).
- Panel: A Panel lays out multiple components in a row, column, or grid.

This is markdown **text**

Fig. 6.113 Holoviews `panel` supports markdown text

cap words

CAP WORDS

Fig. 6.114 The panel `depends` decorator connects callbacks to widgets

- Widget: A Widget provides input controls to add interactive features to your Panel.

The following shows a `pn.panel` object rendering markdown (excluding Math-JaX):

```
>>> pn.panel('### This is markdown **text**')
Markdown(str)
```

You can also have raw HTML as a Panel element,

```
>>> pn.pane.HTML('<marquee width=500><b>Breaking News</b>: some
↪  news.</marquee>')
HTML(str)
```

The main layout mechanism for panes is `pn.Row` and `pn.Column`. There is also `pn.Tabs` and `pn.GridSpec` for more complicated layouts.

```
>>> pn.Column(pn.panel('### This is markdown **text**'),
...     pn.pane.HTML('<marquee width=500><b>Breaking News</b>:
↪  some news.</marquee>'))
Column
    [0] Markdown(str)
    [1] HTML(str)
```

We can use a widget and connect to the panes. One way to do this is with the `@pn.depends` decorator, which connects the inputbox's input string to the `title_text` callback function (Fig. 6.114). Note that you have to hit ENTER to update the text.

```
>>> text_input = pn.widgets.TextInput(value='cap words')

>>> @pn.depends(text_input.param.value)
... def title_text(value):
...     return '## ' + value.upper()
...
>>> app2 = pn.Row(text_input, title_text)
>>> app2
```

Here is a neat autocompletion widget. Note that you have to type at least two characters (Figs. 6.115 and 6.116).

```
>>> autocomplete = pn.widgets.AutocompleteInput(
...     name='Autocomplete Input',
```

Fig. 6.115 Panel
autocompletion widget

Autocomplete Input

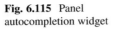

Write something here and <TAB> to complete

Country

United States

1966 .. 1990

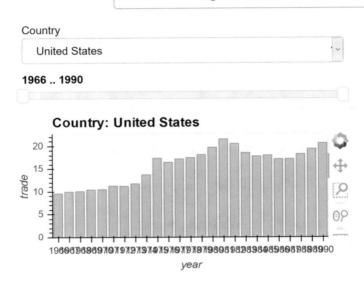

Fig. 6.116 Panel supports dashboards with embedded Holoviews visualizations

```
...          options=economic_data.country.unique().tolist(),
...          placeholder='Write something here and <TAB> to complete')
>>> pn.Column(autocomplete,
...              pn.Spacer(height=50)) # the spacer adds some
↪  vertical space
Column
    [0] AutocompleteInput(name='Autocomplete
↪  Input',options=['United States', ...], placeholder='Write
↪  something h...)
    [1] Spacer(height=50)
```

Once you are satisfied with your app, you can annotate it with `servable()`. In
the Jupyter notebook, this annotation has no effect, but running the `panel serve`
command with the Jupyter notebook (or plain Python file) will create a local Web
server with the annotated dashboard.

We can create a quick dashboard of our `economic_data` by using these elements.
Note the decorator parameters on the `barchart` function. You can set the extremes
of the `IntRangeSlider` and then move the interval by dragging the middle of
the indicated widget.

```
>>> pulldown = (pn.widgets.Select(name='Country',
...                     options=economic_data.country
...                                      .unique()
...                                      .tolist()))
```

```
...                      )
>>> range_slider = pn.widgets.IntRangeSlider(start=1966,
...                                          end=1990,
...                                          step=1,
...                                          value=(1966,1990))

>>> @pn.depends(range_slider.param.value,pulldown.param.value)
... def barchart(interval, country):
...        start,end = interval
...        df=
↪      economic_data.query('country=="%s"'%(country))[['year','trade']]
...        return (df.set_index('year')
...                   .loc[start:end]
...                   .hvplot.bar(x='year',y='trade')
...                   .relabel(f'Country: {country}'))
...
>>> app=pn.Column(pulldown,range_slider,barchart)
>>> app
Column
    [0] Select(name='Country', options=['United States', ...],
    value='United States')
    [1] IntRangeSlider(end=1990, start=1966, value=(1966, 1990),
    value_throttled=(1966, 1990))
    [2] ParamFunction(function)
```

Note that the callback is Python so it requires a backend Python server to update the plot.

6.6 Plotly

Plotly is a web-based visualization library that is easiest to use with `plotly_express`. The key advantage of `plotly_express` over plain `plotly` is that it is far less tedious to create common plots. Let us consider the following dataframe:

```
>>> import pandas as pd
>>> import plotly_express as px
>>> gapminder = px.data.gapminder()
>>> gapminder2007 = gapminder.query('year == 2007')
>>> gapminder2007.head()
        country continent year lifeExp        pop gdpPercap iso_alpha iso_num
11  Afghanistan      Asia 2007   43.83   31889923    974.58       AFG       4
23      Albania    Europe 2007   76.42    3600523   5937.03       ALB       8
35      Algeria    Africa 2007   72.30   33333216   6223.37       DZA      12
47       Angola    Africa 2007   42.73   12420476   4797.23       AGO      24
59    Argentina  Americas 2007   75.32   40301927  12779.38       ARG      32
```

The `scatter` function draws the plot in Fig. 6.117.

```
>>> fig=px.scatter(gapminder2007,
...                x='gdpPercap',
...                y='lifeExp',
...                width=400,height=400)
```

Fig. 6.117 Basic Plotly
scatter plot

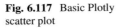

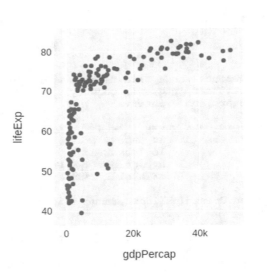

The dataframe's columns that are to be drawn are selected using the x and y
keyword arguments. The width and height keywords specify the size of the plot.
The fig object is a package of Plotly instructions that are passed to the browser
to render using Plotly Javascript functions that include an interactive toolbar of
common graphic functions like zooming, etc.

As with Altair, you can assign dataframe columns to graphic attributes. The
following Fig. 6.118 assigns the categorical continent column to the color in
the figure.

```
>>> fig=px.scatter(gapminder2007,
...                x='gdpPercap',
...                y='lifeExp',
...                color='continent',
...                width=900,height=400)
```

Figure 6.119 assigns the 'pop' dataframe column to the size of the marker in the
figure and also specifies the figure's size with the width and height keyword
arguments. The size and size_max ensure the marker sizes fit neatly in the plot
window.

```
>>> fig=px.scatter(gapminder2007,
...                x='gdpPercap',
...                y='lifeExp',
...                color='continent',
...                size='pop',
...                size_max=60,
...                width=900,height=400)
```

In the browser, hovering the mouse over the data marker will trigger a popup note
with the country name in Fig. 6.120.

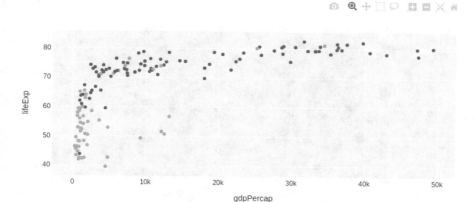

Fig. 6.118 Same scatterplot as Fig. 6.117 but now the respective countries are colored separately

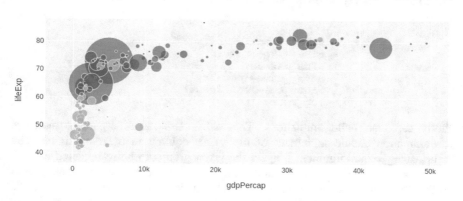

Fig. 6.119 Same as Fig. 6.118 with population value scaling the marker sizes

```
>>> fig=px.scatter(gapminder2007,
...                x='gdpPercap',
...                y='lifeExp',
...                color='continent',
...                size='pop',
...                size_max=60,
...                hover_name='country')
```

Subplots in Matplotlib are known as *facets* in Plotly. By specifying the `facet_col`, the scatter plot can be split into these facets as shown in Fig. 6.121, where `log_x` changes the horizontal scale to logarithmic.

```
>>> fig=px.scatter(gapminder2007,
...                x='gdpPercap',
...                y='lifeExp',
```

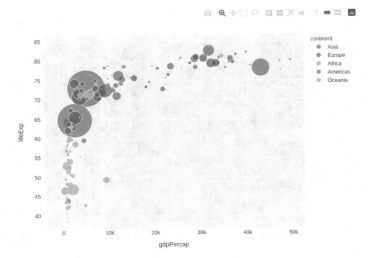

Fig. 6.120 Same as Fig. 6.119 but with hover tool tips on the markers and new plot size

```
...                     color='continent',
...                     size='pop',
...                     size_max=60,
...                     hover_name='country',
...                     facet_col='continent',
...                     log_x=True)
```

Plotly can also build animations. The `animation_frame` argument indicates the animation should increment of the `year` column in the data frame. The `animation_group` argument triggers the given groups to be re-rendered at each animation frame. When building animations, it is important to keep the axes fixed so that the motion of the animation is not confusing. The `range_x` and `range_y` arguments ensure the axes (Fig. 6.122).

```
>>> fig=px.scatter(gapminder,
...                 x='gdpPercap',
...                 y='lifeExp',
...                 size='pop',
...                 size_max=60,
...                 color='continent',
...                 hover_name='country',
...                 log_x=True,
...                 range_x=[100,100_000],
...                 range_y=[25,90],
...                 animation_frame='year',
...                 animation_group='country',
...                 labels=dict(pop='Population',
...                             gdpPercap='GDP per Capita',
...                             lifeExp='Life Expectancy'))
```

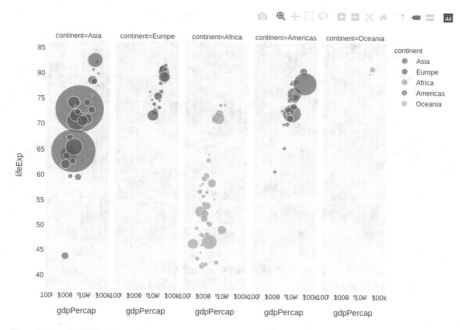

Fig. 6.121 Plotly facets are roughly equivalent to Matplotlib subplots

Common statistical plots are easy with `plotly_express`. Consider the following data:

```
>>> tips = px.data.tips()
>>> tips.head()
   total_bill   tip     sex smoker  day    time  size
0       16.99  1.01  Female     No  Sun  Dinner     2
1       10.34  1.66    Male     No  Sun  Dinner     3
2       21.01  3.50    Male     No  Sun  Dinner     3
3       23.68  3.31    Male     No  Sun  Dinner     2
4       24.59  3.61  Female     No  Sun  Dinner     4
```

The histogram of the sum of the tips partitioned by `smoker` status is shown in Fig. 6.123

```
>>> fig=px.histogram(tips,
...                   x='total_bill',
...                   y='tip',
...                   histfunc='sum',
...                   color='smoker',
...                   width=400, height=300)
```

Everything Plotly needs to render the graphic in the browser is contained in the `fig` object; so to change anything in the plot, you have to change the corresponding item in this object. For example, to change the color of one of the histograms in

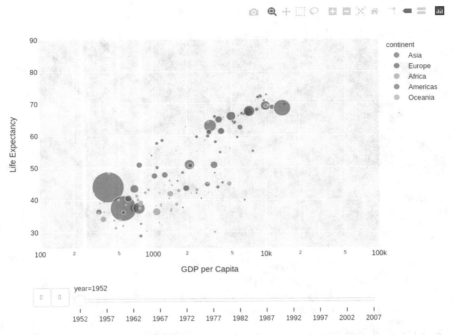

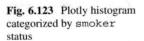

Fig. 6.122 Plotly can animate complex plots with the `animation_frame` keyword argument

Fig. 6.123 Plotly histogram
categorized by `smoker`
status

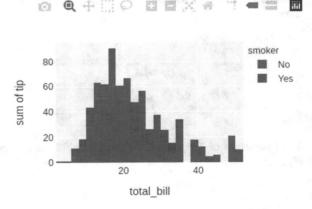

Fig. 6.124 Same as Fig. 6.123 but with color change

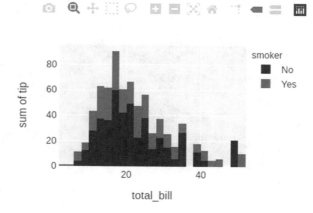

Fig. 6.123, we access and update the `fig.data[0].marker.color` object attribute. The result is shown in Fig. 6.124

```
>>> fig=px.histogram(tips,
...                   x='total_bill',
...                   y='tip',
...                   histfunc='sum',
...                   color='smoker',
...                   width=400, height=300)
>>> # access other properties from variable
>>> fig.data[0].marker.color='purple'
```

Plotly can draw other statistical plots like the notched boxplot shown in Fig. 6.125. The `orientation` keyword argument lays out the boxplots horizontally.

```
>>> fig=px.box(tips,
...            x='total_bill',
...            y='day',
...            orientation='h',
...            color='smoker',
...            notched=True,
...            width=800, height=400,
...            category_orders={'day': ['Thur', 'Fri', 'Sat',
↪ 'Sun']})
```

Violin plots are an alternative way to display one-dimensional distributions (see Fig. 6.126).

```
>>> fig=px.violin(tips,
...               y='tip',
...               x='smoker',
...               color='sex',
...               box=True, points='all',
...               width=600,height=400)
```

Marginal plots can be included using the `marginal_x` and `marginal_y` keyword arguments as shown in Fig. 6.127. The `trendline` keyword argument means that

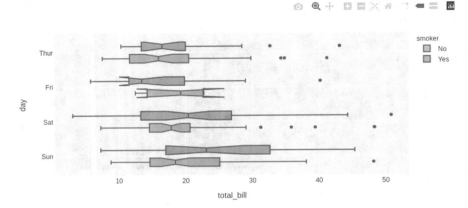

Fig. 6.125 Boxplots can be either horizontally or vertically oriented

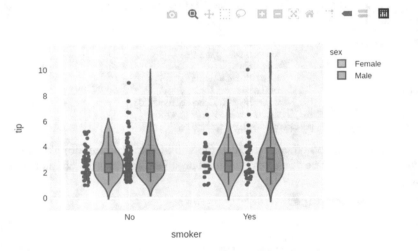

Fig. 6.126 Plotly supports violin plots for one-dimensional probability density function visualizations

the ordinary least-squares fit should be used to draw the trend line in the middle subplot.

```
>>> fig=px.scatter(tips,
...                x='total_bill',
...                y='tip',
...                color='smoker',
...                trendline='ols',
...                marginal_x='violin',
...                marginal_y='box',
...                width=600,height=700)
```

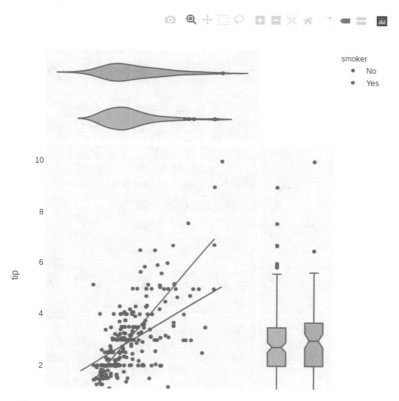

Fig. 6.127 Plots in the margins of a central graphic are possible using the `marginal_x` and `marginal_y` keyword arguments

This short section just scratches the surface of what Plotly is capable of. The main documentation site is the primary resource for emerging graphical types. Plotly express makes it much easier to generate the complex Plotly specifications that are rendered by the Plotly Javascript library, but the full power of Plotly is available using the bare `plotly` namespace.

References

1. Bokeh Development Team. Bokeh: Python library for interactive visualization (2020)

Index

Printed in the United States
by Baker & Taylor Publisher Services